Stokes Nature Guides

by Donald Stokes

A Guide to Nature in Winter
A Guide to Observing Insect Lives
A Guide to Bird Behavior, Volume I

by Donald and Lillian Stokes

A Guide to Bird Behavior, Volume II
A Guide to Bird Behavior, Volume III
A Guide to Enjoying Wildflowers
A Guide to Animal Tracking and Behavior

by Thomas F. Tyning

A Guide to Amphibians and Reptiles

ALSO BY DONALD STOKES

The Natural History of Wild Shrubs and Vines

ALSO BY DONALD AND LILLIAN STOKES

The Bird Feeder Book
The Hummingbird Book
The Complete Birdhouse Book
The Bluebird Book
The Butterfly Book
The Wildflower Book: East of the Rockies
The Wildflower Book: From the Rockies West

A Guide
to
Bird Behavior

Volume III

A Guide to Bird Behavior

Volume III

by Donald W. Stokes and Lillian Q. Stokes

Illustrated by Bob Hines
Display Guides Illustrated by Donald W. Stokes

Little, Brown and Company
Boston   New York   Toronto   London

FIRST EDITION

Chapter-opening illustrations by Bob Hines
Copyright © 1989 by Bob Hines

Library of Congress Cataloging-in-Publication Data
(Rev. for vol. 3)

Stokes, Donald W.
    A guide to bird behavior.

    (Stokes nature guides)
    Spine title: Birds.
    Bibliography: p.
    Contents: —v. 2. In the wild and at your
feeder—v. 3.
    1. Birds—Behavior.   2. Birds—North America—
Behavior.   1. Stokes, Lillian Q.   II. Title.
III. Title: Birds.   IV. Series.   V. Series: Stokes,
Donald W. Stokes nature guides.
QL698.3.S757   1983   598.251′0973   83–14924
ISBN: 0-316-81737-6 (hc)
        0-316-81717-1 (pb)

HC: 10 9 8 7 6 5 4 3 2 1
PB: 10 9 8

MV-NY

*Published simultaneously in Canada
by Little, Brown & Company (Canada) Limited*

PRINTED IN THE UNITED STATES OF AMERICA

Acknowledgments

OUR GREATEST THANKS GO TO THE PEOPLE LISTED IN OUR BIBLIOGRAPHY. Their patient studies of birds are the backbone of this guide. Without them, we could not have introduced readers to the secret lives of birds and the joy of watching their behavior. We also would like to thank, in particular, Lawrence Zeleny, for his helpful suggestions on the bluebird chapter, and Paul Roberts, who helped us compile the information for the hawk-watching charts.

Contents

A Guide to Bird Behavior

Volume III

The Writing of Volume III

ALTHOUGH THE THREE VOLUMES OF *A Guide to Bird Behavior* APPEAR similar, each volume has a purpose quite different from the others. This is most apparent when you look at the choice of species in each volume.

When volume 1 was written there was no thought of any future volumes; in fact, Stokes Nature Guides did not yet exist. At that time, the book was called *A Guide to the Behavior of Common Birds.* A few years later, after the two of us were married and started writing together, we created the idea behind Stokes Nature Guides. The title of the first book was then changed to *A Guide to Bird Behavior,* Volume I, and two additional volumes were planned.

There are many reasons behind the choice of birds in the first volume. The most important is that they are all easy to watch, for they live in open habitats and are not particularly shy of humans. Second, many of them — such as the pigeon, house sparrow, and starling — live in urban areas and were chosen because of this; thus, whether you live in the country or the city, you can participate in behavior-watching. And third, we wanted to include species that represented different bird families. This exposes you to a variety of behaviors and enables you to use your knowledge of one species to help you understand the behavior of other species in the same family.

By the time we started writing volume 2, many people had already told us that they wanted to know about the behavior of more birds and that we had left out their favorites, such as the cardinal, titmouse, nuthatch, oriole, and many others. This led to the choice of birds for volume 2. These are birds that breed throughout most of North America, are common, and nest primarily in country and suburban areas. In a sense, they represent garden and farmland birds.

Volume 3 radically departs from the other volumes in that it contains many species that are hard to watch, and others that are uncommon or even rare. The quality that most of these species share is that people are strongly attracted to them and, when near them, want to know about their lives. In some cases the public interest has been aroused because the species is endangered, as is the case with the osprey, bald eagle, and peregrine falcon. In other cases, the interest is due to the bird's beauty, as with the bluebird and hummingbird, its habit of nesting near humans, as with the purple martin, or its value as a game bird, as with the bobwhite. Other birds, such as the wood duck, common loon, pileated woodpecker, great blue heron, and American woodcock, each hold a special place in the hearts of those who love birds.

We have also included a lot of birds of prey — seven hawks and three owls — because when you cross paths with birds this exciting you really want to get closer to them and know their lives and behavior. Included with the discussion of each hawk is special information on watching it during migration. The present volume also includes an appendix on hawk-watching.

This is the last volume on bird behavior in the Stokes Nature Guides. We hope the new approach to bird-watching presented in these three volumes has enriched your experiences with birds. It has been impossible to include everybody's favorite species, but we hope that most of the birds you enjoy watching are covered. We wish you the best of luck with your behavior-watching.

Yours in Nature,
Don and Lillian Stokes

How to Use This Book

THIS BOOK IS DESIGNED TO HELP YOU DISCOVER AND EXPLORE THE LIVES OF your favorite birds. Begin by glancing over the contents and becoming familiar with the twenty-five species included in this volume. Next, choose a bird that you like, have seen, or think you may see, and browse over the chapter on its life.

For each species, after an initial introduction, there are three sections of information: a behavior calendar, a display guide, and behavior descriptions.

The behavior calendar provides an approximation of the timing of a bird's life cycle and will give you a rough idea of when certain behaviors are most common. Clearly, timing varies widely with latitude. The behavior calendar has been calculated for the middle

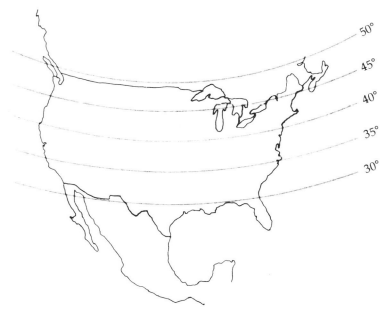

Common Loon
Gavia immer

LOONS HAVE ALWAYS STIRRED THE HEARTS OF HUMANS WITH THEIR mysterious nocturnal behavior and their striking calls, which have been likened to everything from moans of death to insane laughter. Add to this their preference for breeding on isolated northern lakes and it is easy to see why our scanty knowledge of loons has in the past been filled in largely by imagination rather than with facts.

In recent years this has changed. Several researchers have done detailed studies of loons' lives. This has been inspired, in part, by our seeing that loon populations were decreasing due to increasing numbers of summer vacationers on lakes where the birds breed. Strong efforts to protect the loon and create greater public awareness of its needs have been fostered by the North American Loon Fund (see Appendix A for address).

Now all of us can enjoy loons more, and, in fact, you do not even need to see them to start following their behavior. Simply by knowing the meaning of loon "language," you can begin to interpret many of the events of their lives. For example, the so-called "laugh," or Tremelo-call, of the loon is generally given when the birds are alarmed. The long Wail-call is probably a location note that helps a pair keep in contact, and the Yodel-call is used in territorial advertisement and defense.

It is extremely important to remember that loons are easily disturbed during the incubation phase. If you come near the nest, even unintentionally, they will leave through a series of long dives and will not return until you have left the area. The better you know loon behavior, the better you can tell which stage of breeding they are in and whether they are disturbed. Then you can adjust your actions so that the loons can breed in peace and we can continue to enjoy their wild sounds and fascinating behavior.

BEHAVIOR CALENDAR

	TERRITORY	COURTSHIP	NEST-BUILDING	BREEDING	PLUMAGE	SEASONAL MOVEMENT	FLOCK BEHAVIOR
JANUARY	■				■		■
FEBRUARY	■				■		■
MARCH	■				■	■	■
APRIL	■	■				■	■
MAY	■		■			■	
JUNE	■		■	■			
JULY	■			■			
AUGUST				■	■		■
SEPTEMBER				■	■	■	■
OCTOBER					■	■	■
NOVEMBER					■	■	■
DECEMBER	■				■		■

DISPLAY GUIDE

Visual Displays

Bill-Dipping

Male or Female   *Sp Su F W*

Bird rapidly dips its bill in the water and may flick it sideways when bringing it out.

CALL: None

CONTEXT: Done when birds join each other, for example when mates meet or flock members come together. May help reduce aggressive tendencies in the birds and enable them to associate closely. *See* Territory, Courtship, Flock Behavior

Splash-Dive

Male or Female   *Sp Su F W*

Bird dives quickly, creating a large splash with a sharp kick of its legs.

CALL: None

CONTEXT: Done along with Bill-dipping and may have a similar role in reducing aggressive tendencies. Also done with Bill-dipping by a mated pair as part of precopulatory displays. *See* Territory, Courtship

Upright

Male or Female   *Sp Su F W*

Bird raises body upright out of the water while treading with its feet. Wings may or may not be spread out to the side. The white belly plumage is exposed.

CALL: None

CONTEXT: May occur between two birds involved in a territorial encounter. *See* Territory

Crouch-Yodel
Male   *Sp Su F W*

Bird extends its head and neck parallel to the water.

CALL: Yodel-call

CONTEXT: Occurs during aggressive encounters between males at territorial borders. *See* Territory

Vulture-Yodel
Male   *Sp Su F W*

Bird assumes an upright position with wings held out to the side.

CALL: Yodel-call

CONTEXT: Done by males during aggressive encounters at close range at territorial borders or within flocks. *See* Territory, Flock Behavior

V-Flight
Male or Female   *Sp Su*

Bird in flight glides with wings held stiffly in a V over its back. Glide may be in a circle and may continue until the bird lands.

CALL: Tremelo-call

CONTEXT: Occurs most during the breeding season. Its function is not known.

Auditory Displays

Wail-Call
Male or Female   *Sp Su F W*

A long, drawn-out wail, sounding like the howl of a coyote or wolf. May have one or two distinct rises in pitch. May drop down a pitch at the end. Lasts about two seconds.

CONTEXT: May be given when a bird is trying to locate its mate, or when an adult is trying to locate a chick. May also be given by the female during Yodel-calls by the male.

Tremelo-Call
Male or Female   *Sp Su F W*
A short, tremulous or vibrating call on one or more pitches. About a half second long. Referred to as the loon's "laughter."
CONTEXT: Indicates alarm due to disturbance of any kind, even simply the appearance of another loon. Only call that is also made in flight. Call most often combined with other calls, in which case it seems to add alarm or fear to meanings of other calls.

Yodel-Call
Male   *Sp Su F W*
The longest and most complex loon vocalization. Starts with a wail, which is followed by a series of undulations lasting several seconds. Given while the male is doing the Vulture-yodel or Crouch-yodel display.
CONTEXT: Used in territory advertisement and defense. Most common in spring and early summer. Usually heard from dusk to dawn. Used again in winter territories. *See* Territory

Hoot-Call
Male or Female   *Sp Su F W*
A single, short, soft hoot. Can occur at varying pitches.
CONTEXT: Seems to be given as a contact call between mates, and between adults and chicks. May also occur in flocks.

Fledgling-Calls
Chicks peep like chickens as a contact note

with parents. They call almost all the time when very young, less as they mature. Chicks have high-pitched voices through the summer, and gradually acquire an adult voice in fall and winter.

BEHAVIOR DESCRIPTIONS

Territory

Summer Territory
Type: Mating, nesting, feeding
Size: 60 – 200 acres
Main behavior: Yodel-call, Crouch-yodel, Vulture-yodel
Duration of defense: From first arrival until the fledgling phase

Loons arrive on their breeding lakes often as soon as a few days after the ice breaks up. Pairs return to breeding spots used in previous years and immediately begin to establish a territory. Territories are formed on and around lakes that have open water deep enough for diving and that have either islands or other shoreline spots suitable for nesting.

If a lake is under a hundred acres in size then just one pair may nest on it. On larger lakes there may be two or more pairs. In these cases, territories are usually defined in relation to landforms around the lake, such as bays, inlets, or peninsulas. Territory sizes vary depending upon conditions for feeding and nesting. They range from sixty to two hundred acres. Throughout the breeding season, on larger lakes, there may also be neutral areas between territories where loons may be seen together without displaying aggression.

Most territorial advertisement and defense is done through the Crouch-yodel, a visual display accompanied by the Yodel-call and done only by the male. The sight of another loon near the territory stimulates the first loon to do the Crouch-yodel, and the intruder or neighbor may answer, sometimes even overlapping the call of the other bird with its own. The encounter stops after a few Crouch-yodels are given and one male and his mate begin to make dives and move away.

If the two males get very close to each other they may do the Vulture-yodel, which seems to be a more intense version of the Crouch-yodel.

When loons meet at borders the interaction is not always strictly aggressive; a set of displays may occur that appease aggressive tendencies. These may be done by neighbors that know each other and agree on a territorial border. These include Bill-dipping, Splash-dives, and circling around each other. Another behavior included in these nonaggressive encounters is "rushing" — when one bird flaps its wings while darting over the water.

Although territories are defended from the time of the pair's first arrival, the defense intensifies gradually up to the time the young hatch and are out on the water. During the early stages of egg-laying and incubation, the pair remain more or less in a restricted area right around the nest and do not chase other birds on the periphery of their territories. But, as the time when the young will hatch nears, they defend a larger and larger area, which brings them into more conflict with neighbors; thus more Yodel-calls are heard at this time. In the following weeks, as the young become better swimmers and divers, the family may roam more widely and not defend the territory as strictly. By late summer there is very little territorial defense.

Sometimes, early in the breeding season, Yodel-calls may be given in succession by a number of birds on a lake or even on separate lakes. The females usually give the Wail-call as the males yodel. These choruses may be a kind of territorial advertisement.

Some pairs do not successfully breed. During the incubation period these pairs are easy to single out because both members are seen as they patrol the boundaries of their territories. Normally, only one member of a pair would be seen at this time, for the other is incubating. Because of their frequent patrolling, these pairs may engage in more aggressive encounters with neighbors than a successfully breeding pair. Pairs that do not successfully breed may also abandon their territories earlier than successful breeders. *See* Flock Behavior

A lone, unmated bird may be able to move unchallenged through a breeding pair's territory as long as it stays away from the nest area and is not too obtrusive. These single birds may spend more time on territories of unsuccessful pairs and may be better tolerated by them.

The Tremelo-call is used at times of disturbance such as when you are on the territory.

Winter Territory

Type: Feeding
Size: 10–20 acres
Main behavior: Yodel-call, Tremelo-call, Upright
Duration of defense: For daylight hours during winter

Wintering loons feed along coastlines. During the day they stay in individual feeding territories, which they defend through the displays of the Yodel-call, Tremelo-call, and Upright posture. These territories are about ten to twenty acres in size and often in bays or inlets. At sunset, the birds stop feeding and join into rafts of several birds in the deepest parts of coves. The next day they disperse again into their territories.

Courtship

Main behavior: Bill-dipping, Splash-dive
Duration: Not known

When pair formation takes place among common loons is not known. The birds seem to arrive at their breeding territory already paired in spring. This means that it either takes place in late summer, fall, or winter. Some evidence suggests that interactions between the sexes occur in the mid- to late-summer flocks. In these flocks there are aggressive interactions between males involving the Vulture-yodel display. These instances may represent a mated male fighting off another male that is challenging him for his mate. It also may be that males are thus establishing dominance in these flocks and thereby gaining access to a female.

A few displays are seen between mates in the spring when the birds arrive on their territories. Two of the most common are Bill-dipping and Splash-dives. Other possible behaviors at this time include the birds' peering under the water, rolling their heads over their backs, and giving slight upward flicks of their bills. These may occur whenever the two birds come together after being apart.

At other times, when just Bill-dipping and Splash-dives occur, they function as precopulatory displays. After giving them, the pair swim off toward shore and the female walks just out of the water and sits down. The male follows, immediately copulates with her, and then leaves soon after. The female usually waits a few minutes before rejoining him. Copulation and the associated displays can be seen from the arrival of the pair on the breeding territory up until egg-laying. They often occur in the pre-dawn or early-morning hours.

Several other displays are seen among loons but their function is still not known. In the past, various writers have associated them with courtship or territory formation and defense, but neither interpretation has been sufficiently proven. In one set of behaviors two or more loons may approach one another, do Bill-dipping, Splash-dives, Upright, and circle around each other. Then one, both, or all may take off across the water, beating their wings against the surface. The Tremelo-call is often given with this action.

Another display occurs in flight. It is especially common in mid- to late summer. The birds glide with wings held motionless over their backs in a stiff V. The flights may be undertaken by one to several birds at once, sometimes in an arc or near-circle, and sometimes completed by landing on the water. Again, their significance is not known.

Nest-Building

Placement: On land, generally an island or peninsula, at the water's edge
Size: About 2 feet in diameter
Materials: Surrounding vegetation, such as reeds, grasses, water plants, moss

Within a few days of arriving on their territory, the pair may cruise around the shoreline looking for suitable nesting spots. Most often islands are preferred; lacking these in their territory, the birds may choose peninsulas or headlands projecting into the lake. When islands are chosen, the birds seem to prefer those that are small — less than two acres in size. The birds often build in the same general area that they used in previous years.

The nests are usually sheltered by surrounding vegetation and upon first being built are right at the water's edge. Subsequent rainfall or drought may affect their distance from the water after this. Nests are built of earth, grasses, moss, or dense floating vegetation.

Little in the way of a nest is built before the eggs are laid. The nest at this time is usually only a little depression in the substrate with a small amount of material collected around it. The majority of the material is added during incubation, for when settling down, the incubating bird usually draws in vegetation from the vicinity, placing it around its body. This is done by both male and female and results in a cleared space around the nest.

Breeding

Eggs: 2, occasionally 1. Brown to olive
Incubation: 29 days, by both male and female
Nestling phase: 1 day
Fledgling phase: 2–3 months
Broods: 1

Egg-Laying and Incubation

The eggs are laid one to two days apart and incubation may begin after the first egg is laid. During incubation the pair stay close together and within a smaller area of their territory, right near the nest. They are generally quiet and very alert to any disturbances.

Both adults incubate the eggs, the female probably doing the greater share. How much the male does varies with individual pairs. Stints on the nest vary from less than an hour's length to most of the day. When ready to take over, the bird off the nest approaches the nest, at which time the other leaves and swims off.

Before leaving, it may do some shuffling of material around the edge of the nest and even continue these types of movements when out on the open water. The bird starting incubation may turn the eggs with its bill, then sit over the eggs facing the water and do some rearranging of nesting material.

Nestling Phase

The young hatch up to a day apart. The first one to hatch stays on or near the nest until the other has hatched and both have dried. Then they leave the nest with the parents and return to shore only occasionally when the young are cold and need to be warmed by being sheltered under the wing of the parent. From the time they hatch well into their fledgling phase, they are fed whole food by the parents.

Fledgling Phase

The young loons follow the parents about and always are attended by one or both parents when out on the open water. The only time they do not stay with the parents is when danger approaches. At these times, the young seek cover along the shore and the parents dive underwater and surface away from the young. They give the Tremelo-call and then dive and surface further away. This appears to be a distraction display that is supposed to lead the predator away. When danger is far enough away, the parents will return quietly to the young through a series of long dives.

After the young are two to three weeks old, they may not head to shore during danger; instead, they dive and attempt to follow the parents. When they are older than this they do the same kind of dives as the parents, moving away from the predator.

The young loon has the wonderful habit of riding on the parent's back while in the water. This may help keep the young warm while they are still small. The adult may lower itself in the water to make it easier for the young to climb aboard. It may also shed the chick by sinking down and leaving the chick afloat. The chick rides on the parent's back as much as half the time during the first four days after fledging and less from then on until they stop, by the second or third week.

The parents feed the young about every hour. One parent may stay with them while the other goes off to catch food items, which are then fed directly to the young. Feeding may last from a few minutes to more than a half hour. The young are fed by the parents for two to three months or longer, even though they can feed on their own before this.

When two days old, the chicks can dive very short distances. This ability continues to improve rapidly and by the time they are a week old they can dive a distance of fifty feet and to a depth of ten feet. At two to three months the young can fly and become independent of the parents.

Plumage

DISTINGUISHING THE SEXES The male and female are identical in plumage and also difficult to distinguish through behavior. Some people suggest that when a mated pair are interacting the male points his bill slightly above the horizontal position and the female points hers slightly below the horizontal position. The male is the only one to give the Yodel-call.

DISTINGUISHING JUVENILES FROM ADULTS The juveniles are almost all black in their first summer, with light areas on their throat and belly. In their winter plumage they look similar to the adults.

MOLTS A complete molt in fall and a partial molt in late winter give the loon two distinct plumages. The winter plumage is all dark gray except for the lighter throat and belly. The summer plumage is mostly dark with white spots on the back and a white and black neck ring.

Seasonal Movement

Loons migrate by day along the coast, farther out to sea, or inland. They migrate singly or in groups of from two to fifteen. Coastal

Great Blue Heron
Ardea herodias

FOR MOST OF US, SIGHTINGS OF GREAT BLUE HERONS ARE CONFINED TO A glimpse of the bird as it flies slowly and steadily overhead, wings arching gracefully down with each beat, neck bent back, and feet trailing behind. At other times we see it on its feeding grounds, standing motionless and staring intently into shallow water, or wading with measured steps as it searches for prey.

The bird's nesting behavior is not often witnessed since, rather than breeding in dispersed areas, it nests in small, concentrated colonies in fairly isolated areas. An inquiry to the local nature center or birding group, however, can lead you to the location of the nearest colony. Colonies should be visited with care and only so long as no disturbance of the birds results. Even one visit will reveal a wealth of behavior.

But even if you cannot arrange to see breeding behavior, great blue herons are fascinating to watch on their feeding grounds. Where many herons feed together, individuals may form temporary territories that they defend with aggressive displays, calls, and chases. And since the birds are so large, these can be very dramatic.

Also, as you watch herons away from the breeding ground, take note of whether the birds you see are juveniles or adults. It is easy to tell by looking at their heads. The juveniles have an all-dark head; the adults have a white crown. At various times you will see more adults or more juveniles. These differences reflect other social changes in the birds' lives, such as the end of the breeding season.

BEHAVIOR CALENDAR

	TERRITORY	COURTSHIP	NEST-BUILDING	BREEDING	PLUMAGE	SEASONAL MOVEMENT	FLOCK BEHAVIOR
JANUARY	■						■
FEBRUARY	■				■	■	■
MARCH	■	■			■	■	
APRIL	■	■	■	■		■	
MAY	■		■	■			
JUNE	■			■			
JULY	■			■			
AUGUST	■				■		
SEPTEMBER	■				■	■	
OCTOBER	■					■	
NOVEMBER	■					■	■
DECEMBER	■						■

DISPLAY GUIDE

Visual Displays

Upright

Male or Female   *Sp Su F W*

Bird stands tall with head and neck at a forty-five-degree angle.

CALL: None

CONTEXT: Most common during feeding, when each bird defends its own territory or personal space. A bird does this display as a first warning to an intruder. May also occur at the nest. *See* Territory

Upright-Head-Down

Male or Female   *Sp Su F W*

Bird stands tall with neck straight out at a forty-five-degree angle but head pointed down, with crest, neck, and back feathers all raised. The bird may walk in a stiff manner toward an intruder.

CALL: Sometimes Rok-rok-call

CONTEXT: Done in response to an intruder on the feeding grounds or, occasionally, at the nest. A more aggressive display than the Upright. *See* Territory

Bent-Neck

Male or Female   *Sp Su F W*

Body is horizontal and neck is bent with head pulled back as if ready to strike out. Bird may walk rapidly toward intruder with wings slightly open.

CALL: Rok-rok-call

CONTEXT: The most aggressive display; attack is imminent. Done most on the feeding grounds or at the nest. *See* Territory

Head-Down

Male or Female   Sp Su F W

Legs are bent, neck and head are extended down below the body level. Crest, neck, and back feathers may be raised.

CALL: Bill-clap

CONTEXT: Done as a territorial display on the nest when defending it from other herons. Also may be done together by a mated pair on the nest, their necks crossing occasionally during the display. *See* Territory, Courtship

Stretch

Male or Female   Sp Su

Bird points bill and neck directly up and then lowers head down toward back while bill is still pointing up.

CALL: Howl

CONTEXT: Done between members of a pair at the nest. *See* Courtship

Swaying

Male or Female   Sp Su

Male and female lock tips of their bills and move their heads back and forth together.

CALL: None

CONTEXT: Done as part of courtship on or near the nest. *See* Courtship

Auditory Displays

Rok-Rok-Call

Male or Female   Sp Su F W

A short, harsh, guttural call usually repeated twice.

CONTEXT: Usually given in aggressive situa-

tions along with a visual display or while flying after an intruder.

Frahnk-Call
Male or Female   *Sp Su F W*
A harsh, drawn-out guttural call.
CONTEXT: Given during aggressive interac-
tions, such as chases in a feeding territory.

Bill-Snap
Male or Female   *Sp Su F W*
A sharp clacking sound made as the bill is
snapped shut.
CONTEXT: Part of the Head-down display.

Howl
Male or Female   *Sp Su*
A drawn-out crooning sound.
CONTEXT: Accompanies the Stretch display.

Nestling-Calls
These are repeated sounds such as "kak-kak-
kak-kak"; they are harsh and clacking and
can be heard from quite a distance.

BEHAVIOR DESCRIPTIONS

Territory

Type: Mating, nesting
Size: Immediate area around nest
Main behavior: Head-down, Bill-clap
Duration of defense: From start of breeding until the fledgling phase

During the breeding season great blue herons defend the immediate area around their nest by giving aggressive displays. The
most common of these is the Head-down display, which is accom-

panied by the Bill-clap sound. Three other aggressive displays may occur at the nest but are more often seen at feeding territories; these are the Upright, Upright-head-down, and Bent-neck displays.

At any time of year, great blue herons may establish temporary feeding territories. These can be from a few to several hundred yards in diameter. The size of the territory and even its very existence depends on the availability of food. Only in times of relative scarcity are these territories vigorously defended. Some researchers believe that juvenile birds do not form feeding territories but, instead, feed in loose flocks.

Defense of feeding territories is commonly seen and involves aerial chases, Frahnk-calls, and aggressive displays, such as Upright, Bill-down-upright, Bent-neck. Fighting rarely occurs, but when it does it can be violent, with one bird landing on the back of the other and either bird stabbing the other with its bill.

Courtship

Main behavior: Head-down, Stretch, Swaying, Allo-preening
Duration: From arrival on breeding ground through nestling phase

Great blue herons probably do not breed until their second or third year, when they acquire their full adult plumage.

Courtship starts when the birds arrive on the breeding ground and continues through the nestling phase. Courtship has at least four associated displays. The Head-down, which is part of territorial defense, also may be given between mated birds early in the breeding season. They may both do this display over the edge of the nest and even cross necks in the process.

Another courtship display occurs when the male brings nesting material to the female. She does the Stretch display accompanied by the Howl-call. The male may do the Stretch-display at other times when the female arrives at the nest after being away.

A third display associated with courtship is Swaying (when mates lock bill tips and move their heads in unison back and forth). This is usually done on or near the nest as well.

And finally there is Allo-preening, when a bird preens or rubs its bill over the feathers of its mate's head, neck, and back. How all of these elements fit into the total picture of great blue heron courtship is not yet known.

Copulation is preceded by one or more of the displays listed above. It occurs on the nest or on a nearby branch with the female crouching down and the male stepping on her back and grasping her neck feathers in his bill.

Nest-Building

Placement: In tall trees or, in the absence of tall trees, lower in shrubbery; rarely on the ground
Size: 2 to 3½ feet outside diameter
Materials: Small branches, twigs, tree leaves, or grasses

Great blue herons nest in colonies with other great blue herons and other species of herons. These rookeries are usually in isolated spots, well away from human settlements and from potential disturbance. Great blue herons prefer to nest in tall trees, usually at the top on vertical branches. They often nest in trees on islands or in trees with water around the base, possibly for added protection from ground predators, such as raccoons.

Rookeries may be used for decades and have many nests, even in excess of the number of pairs breeding there at any one time. Nests are used in subsequent years and become substantial and strong, surviving gale-force winds.

Upon arrival at a rookery in spring, the birds do not start building right away but remain perched in the trees, preening, resting, and making trips to feeding spots. After a few days they start to claim nest sites and defend them against intrusion from other herons.

Sticks are gathered from the ground, from trees, and from old or even active nests. The twigs chosen are a foot or more long and about a half inch in diameter. The male generally brings the material to the female, who greets him with the Stretch-display, then takes the twig and places it in the nest. The nest may take from a few days to a few weeks to complete. Some nest-building may continue through incubation.

When birds are using a nest from a previous year, they just renovate it with a few new twigs. Whether the nest is old or new, the birds line it with finer twigs, grasses, and/or leaves.

In older rookeries, the vegetation at the base of the nest trees and the nest trees themselves may be dead. This probably results from the buildup of droppings excreted by the birds.

Breeding

Eggs: 3–5. Bluish green to pale olive
Incubation: Approximately 28 days, by male and female
Nestling phase: 7–8 weeks
Fledgling phase: 2–3 weeks
Broods: 1

Egg-Laying and Incubation

Laying of the first egg occurs several days after the nest is completed. Three to five eggs are laid at intervals of at least two days. Incubation may start before the laying of the last egg, which results in the young hatching over a period of several days.

There seems to be little ceremony as the birds exchange positions on the nest. The arriving bird may give a low gurgling sound as it approaches. The incubating bird gets up, stands on the rim of the nest, and flies off as the other arrives. Occasionally, the arriving bird may bring twigs and give them to the incubating bird, who places them in the nest.

The arriving bird carefully settles over the eggs, sometimes turning them some. When incubating, a great blue heron is very still and quiet. In fact, this is the quietest time at the rookery. The rookery will not be entirely quiet, however, because other pairs may be in different stages of breeding since there is generally little breeding synchrony in a great blue heron rookery.

Nestling Phase

The young hatch over a period of up to seven days. This means that there can be large discrepancies among the sizes of the young. At first the young are brooded almost constantly by the parents, who, in contrast to their incubating behavior, are more fidgety and active while sitting over the young.

The newborn young are fairly helpless and can only crawl around in the nest. Both parents bring them food. The parents at first stand on the side of the nest as they begin to regurgitate, then bend down and carefully place bits of food into the mouths of the young. Generally, after feeding them in these early stages the adult broods them.

As the young get older they become more active and begin to make their incessant "kak-kak-kak" calls when they see the parents in the vicinity. At this later stage, the parents regurgitate food into their own bills and the young poke their bills into that of the parent when feeding. In the last stages of nestling life the parents may just regurgitate the food into the nest. The young then fight over it.

In many cases one or more members of a brood do not survive. The ones that die are most often the smallest ones, those that hatched later than the others. They have trouble competing for food and are believed to die of starvation. This is a natural way to reduce the number in the brood when there is not enough food.

The young attempt to defecate over the edge of the nest but do not always make it all the way. As a result the edge of the nest may be stained with whitewash.

A defensive reaction of the young when there is danger under the nest is to regurgitate food down on the intruder — an unwelcome surprise for any observer, human or otherwise.

Fledgling Phase

The young leave the nest when they are seven to eight weeks old, at which time they are about the size of the parents. Before that, they may walk out on branches near the nest to exercise their wings. It is not uncommon for one or more young birds in a colony to fall out of the nest and to be killed in doing so.

The young remain with the parents for two to three weeks after leaving the nest and are fed by them during this time. After this they go off on their own and feed in areas with other young.

Plumage

DISTINGUISHING THE SEXES There is no way to distinguish male from female by appearance. There are also very few behavioral clues to the distinction. One is nest-building, which is done primarily by the female at the nest. Another, of course, is the relative position during copulation — the female under the male.

DISTINGUISHING JUVENILES FROM ADULTS Juvenile great blue herons do not look exactly like the adults until after their third fall. Before that there are several ways to tell them apart. One is the lack of a white crown on the juvenile's head (its head is all black). The other is the lack of black marks on the front edge of the juvenile's wings (what looks like the shoulder when the wings are folded). These gradually develop as the bird matures.

MOLTS The birds have a complete molt of all feathers in late summer and early fall and a second partial molt of body feathers in late winter and early spring.

Seasonal Movement

After breeding and when the young can fly, great blue herons may disperse in all directions, with the young birds traveling the farthest in many cases. At this time, late summer, they may even move into areas where they are never seen at any other time of year, for instance farther north than they breed. Following a few weeks of feeding in these new areas, the birds start to migrate south.

Fall migration occurs from mid-September through October. Birds in the Northeast move down the coast; birds in the Midwest may move down to the Gulf of Mexico or to central and western Texas or Mexico. Birds in the South and on the West Coast generally are year-round residents and move only locally to take advantage of food sources.

Fall migration occurs mainly during the day but may also occur to some extent at night. The birds migrate singly or in small flocks of from a few to a dozen or more.

Spring migration starts in February and continues into May for the birds that go the farthest north.

Flock Behavior

In the winter, great blue herons may join other local herons as they go to a communal roost, flying there each night and then flying away every day to favored feeding sites.

Wood Duck
Aix sponsa

THE WOOD DUCK IS THE ONLY NATIVE SPECIES OF PERCHING DUCKS IN North America. Perching ducks are characterized by their having well-developed claws, long tails, and a great deal of iridescence on the wings. They also tend to live in tree holes.

As European settlers started to clear the land in eastern North America, fewer tree holes remained available for wood ducks. This led to a gradual decline in wood duck populations. By the early 1900s wood ducks were almost extinct.

Conservation efforts were begun and protection from hunting led to the gradual increase in populations. However, the hurricane of 1938 blew down many of the wood duck's nest trees in the Northeast. Following this, a program involving setting up artificial nest boxes was started in Massachusetts at Great Meadows National Wildlife Refuge. It proved to be successful and spread to other areas of the country. At present, increased numbers of natural cavities, due to maturing forests, and nest-box programs sponsored by sportsmen's and wildlife organizations are keeping wood duck populations at a healthy level.

Nest boxes afford the behavior-watcher great opportunities to observe breeding behavior in public wildlife areas. Wood duck activities can be seen throughout the year, but of particular interest are the courtship behaviors of fall and spring. They involve many complex displays that are exciting to watch.

BEHAVIOR CALENDAR

	TERRITORY	COURTSHIP	NEST-BUILDING	BREEDING	PLUMAGE	SEASONAL MOVEMENT	FLOCK BEHAVIOR
JANUARY	▨					▨	
FEBRUARY	▨				▨	▨	
MARCH	▨	▨			▨		
APRIL	▨	▨	▨		▨		
MAY			▨				
JUNE			▨	▨			
JULY				▨		▨	
AUGUST				▨		▨	
SEPTEMBER	▨			▨		▨	
OCTOBER	▨				▨	▨	
NOVEMBER	▨				▨	▨	
DECEMBER	▨					▨	

DISPLAY GUIDE

Visual Displays

Turning-the-Back-of-the-Head
Male　　　　　　　　　　　　　*F W Sp*

Wings and tail are slightly raised as male
swims away from female with the back of his
head turned toward her. She usually follows.

CALL: None

CONTEXT: Done soon after pair formation; may
strengthen pair bonds. Done in response to
female's Inciting display. Most common male
display. *See* Courtship

Inciting
Female (or Male)　　　　　　　　*F W Sp*

The female repeatedly and rapidly flicks her
bill back over her shoulder.

CALL: None

CONTEXT: Most common female courtship display. Done near mate and directed at an
intruding male. Male mate usually does
Turning-the-back-of-the-head as the two
swim away from the intruder. Occasionally
done by a paired male upon the approach of
an intruding male. *See* Courtship

Bill-Jerk
Male or Female　　　　　　　　　*F W Sp*

Bird in normal swimming position rapidly
flicks its bill up, exposing the white under its
chin. May be repeated.

CALL: None

CONTEXT: Given as a greeting between paired
birds. Also given repeatedly and synchro-

nously as a precopulatory display by male and female.

Display-Shake
Male   *F W Sp*

Bird starts display by being still for several seconds with head tucked in and crest raised. Then it suddenly arches its neck and, momentarily, lifts its breast out of the water, all the while keeping its head pointing down. Male is usually within two feet of female and faces her side during display.

CALL: None

CONTEXT: Done toward the female when the pair are alone or when an intruding male is near his mate. May help strengthen pair bond. *See* Courtship

Preen-Behind-the-Wing
Male or Female   *F W Sp*

Bird starts with head tucked in and crest raised. Bird dips head down as if to drink, then flicks his bill toward the female, sometimes splashing drops of water at her. Then he reaches back under his wing on the side of the female as he fans the wing, showing the white underneath. Male usually oriented in front of and perpendicular to female.

CALL: None

CONTEXT: Done most in spring by mated birds. Only rarely seen in fall. May be a pair-maintenance display. Occasionally done by female. *See* Courtship

Wing-and-Tail-Flash

Male   *F W Sp*

Bird starts display with head tucked back on body and crest partly erect. Then head, wings, and tail are suddenly lifted up and crest is fully erect. Within a second, the bird returns to its normal position. During display, male is usually oriented in front of and perpendicular to the female. After displaying, bird may wag tail and swim a few feet away from female.

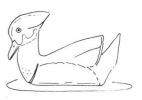

CALL: A short, guttural sound is made during display.

CONTEXT: Most likely a pair-maintenance display. *See* Courtship

Burp

Male   *F W Sp*

The head and crest are raised, the bill is flicked laterally toward the female, and then head and crest are lowered. May be repeated several times in succession.

CALL: Pfit-call during head flicking

CONTEXT: Given as part of aggressive displaying when birds are in groups. *See* Courtship

Bill-Jab

Male or Female   *F W Sp*

Bill is repeatedly jabbed straight down into the water.

CALL: Bill-jab-call

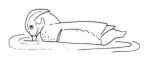

CONTEXT: Aggressive display given by birds in displaying groups. Stimulates the same display in other wood ducks. *See* Courtship

Auditory Displays

Oo-Eek-Call
Female   *F W Sp Su*

The familiar, drawn-out call that most people recognize as the "wood duck's call." Given only by the female. Rises in pitch near the end.

CONTEXT: Given by female as she flies or swims into an area, or as she approaches a group of males. May advertise her presence to males. A shortened version may serve as an alarm note.

Pfit-Call
Male   *F W Sp*

A short, high-pitched whistle. Sounds much like the written "pfit."

CONTEXT: Given by the male during the Burp display. Given most in morning and evening when in courtship groups. *See* Courtship

Ji-Ihb-Call
Male   *F W Sp Su*

A drawn-out, reedy whistle accented at the end. Soft and heard only at close range.

CONTEXT: Given by male at dawn and twilight as birds enter the roost or settle down for the night. Also heard during display in courtship groups.

Bill-Jab-Call
Male or Female   *F W Sp Su*

A soft, rapidly repeated note given during Bill-jab display. In male, sounds like "jibjib-

jibjib," in female, sounds like "dihdihdih-
dih." Best recognized by its accompanying
display.

CONTEXT: Given by male or female during aggressive encounters associated with courtship
display. *See* Courtship

Coquette-Call
Female   *F W Sp*
A short descending note.
CONTEXT: Given during courtship displays.

Te-Te-Te-Call
Female   *Sp*
A rapidly repeated note. Best distinguished
from the similar Bill-jab-call by its not being
accompanied by the Bill-jab display.
CONTEXT: Given during nest-searching. *See*
Nest-Building

BEHAVIOR DESCRIPTIONS

Territory

There are no reports of wood ducks defending territories.

Courtship

*Main behavior: Bill-jerk, Inciting, Turning-the-back-of-the-head,
chases*
Duration: Fall through spring

Courtship in wood ducks has roughly three parts: group displays and aggression, pair formation, and pair maintenance.

Courtship starts in mid-September when most of the adult males have acquired their colorful breeding plumage. Small groups composed of more males than females and totaling about ten birds begin to gather together. Interactions start with several males swimming around a female, their heads often tucked down against their bodies.

They then do Bill-jerk and Whistle-call displays while the females may do the Coquette-call. Both males and females may do the Bill-jab display at this time as well. Displaying increases in intensity and a female may do Inciting. This usually leads to a fight between several males in which they lower their heads and chase after one another with much splashing. The fight usually ends the sequence, and the males that fought may then bathe.

Following a display sequence such as this, which may last from one to twenty minutes, all the birds in the group loaf, preen, and feed. A new sequence may occur soon or over an hour later. Most displaying occurs in the early morning or late afternoon. Wood ducks are generally silent except when displaying, so listen for the calls to locate displaying groups.

These group displays may function to establish a dominance hierarchy among males. The most dominant male or males are the first ones in a group to closely approach females. Pair formation seems to be accomplished with few displays. A female will soon accept only one or two males near her, showing aggression to the others. The chosen males will also keep other males away from the female through aggression and chases. Eventually the female lets only one male stay near, and he may come close enough to preen the feathers on her head.

Once paired, male and female stay together and often avoid displaying groups of wood ducks. If they are separated briefly, the pair may do the Bill-jerk display as a sort of greeting when they rejoin. If another male approaches, the female will do Inciting directed at the strange male, and her mate will do Turning-the-back-of-the-head as the pair swim away from the intruder. The combination of these last two displays is the most common courtship behavior of wood ducks and a good indication that pair formation has occurred.

For the first few weeks of courtship, males continue to compete over females, so pairs may be unstable, lasting for as little as ten minutes. The more elaborate courtship displays are done by paired males near their mate. They are sometimes done in response to an intruding male, and, in other cases, done after the paired male has chased other males away. The most common of these more elaborate displays are: Display-shake, Preen-behind-the-wing, Wing-and-tail-flash, and Burp. These are all described in the Display Guide.

By mid-winter the majority of females seem to be paired. Paired birds remain together and display less than unpaired birds, which tend to stay in groups. More displaying occurs in fall than in spring.

The pair is usually alone when they copulate, often because the male has chased other birds away. Bill-jerk and drinking are done by both male and female in a synchronized way before copulation. Gradually, the female assumes a prone posture, the male approaches from behind her, swims onto her back, grabs the nape of her neck in his bill, and they mate. The male then gets off, releases her from his bill, and swims around to face her while she bathes; he may also do Wing-and-tail-flash and Turning-the-back-of-the-head at this time. Copulation is common in the spring. When it occurs in fall and winter it is not for fertilization but more likely for pair maintenance.

Nest-Building

Placement: In tree holes or human-made boxes
Size: Entrance, 3½ inches or greater; inside, 11 inches in diameter or greater; cavity depth, one foot or more
Materials: Lined with feathers from the female's breast

Paired wood ducks start to look for nesting sites in spring. Most females return to their previous breeding area for successive broods. Yearlings also tend to return to near where they were born to look for nesting sites.

Nest-site searching is done mostly in the morning. Pairs of ducks fly into an area and perch in trees where there are possible hollows in which to nest. The female does most of the actual looking inside holes, while the male remains perched nearby. She may just perch at the hole's entrance or may actually go in.

The birds may use old woodpecker homes, such as those of flickers or pileated woodpeckers. Or they may use naturally rotted holes or human-made boxes. The nest hole can be from 3½ to 60 feet high and over land or water. The female needs at least a 3½-inch-diameter hole to enter. The cavity can be from 1 to 8 feet deep. Interior diameter needs to be about 11 inches.

Starlings may compete with wood ducks for nest holes. Ducks looking for holes occasionally get stuck in house chimneys.

Once she has chosen a nest site, the female adds down from her body to the nest cavity. In natural cavities there are also often wood chips in the bottom of the nest.

Breeding

Eggs: 11–14. White to pale buff
Incubation: 27–30 days (usually 30), by female
Nestling phase: 1 day
Fledgling phase: 5 weeks
Broods: 1, occasionally 2 in warm climates

Egg-Laying and Incubation

Before laying her first egg, the female usually creates a cup-shaped depression in the litter on the bottom of the box. As soon as the first egg is laid, she covers it with litter, making a small cone in the center of the nest.

On days when egg-laying occurs, both male and female approach the nest early in the morning. The female enters the nest hole while the male waits nearby. She lays one egg per day until the clutch is complete. During the laying of the first eggs, the female may spend only a few minutes on the nest, cover the eggs with litter, and then leave. During subsequent layings, she spends more time — up to an hour or more — on the nest.

The female also plucks down from her breast to cover the eggs and keep them warm. This usually starts around the time the sixth to eighth egg is laid. By the end of egg-laying there is a considerable amount of down in the nest.

While most female wood ducks lay about fourteen eggs in their own nest, many nests contain fifteen to twenty eggs or more. These extra eggs are the result of what has been termed "dump nesting" or "egg-dumping," which is when a female lays eggs in another's nest. It is an example of intraspecific parasitism. In one study, 37 percent of the eggs found in nests in a given area were the result of egg-dumping. Thus, it is a common practice among female wood ducks.

Studies of egg-dumping females have shown that they usually arrive at nests with their mates in the morning, when the host is away; they lay in several nests, and spend very little time in the nest after laying. Some egg-dumpers never have a nest of their own; others dump eggs and then go on to have their own nest. Females with nests do not react in any negative way to dumped eggs; they just incubate them and raise the young as if they are their own.

Scientists are still puzzling over the benefits of this strategy. One obvious advantage to the egg-dumper and her mate is that they do not have "all of their eggs in one basket," so to speak.

When incubation starts for wood ducks is hard to determine. The female may spend the night with the eggs during the days when the last four are being laid. Also, the eggs are covered with down. Both of these actions could effect some incubation.

In most cases incubation probably does not start until the last egg is laid. Once incubating, the female leaves the nest only twice a day, in the early morning, sometimes before dawn, and in the late afternoon. During these times she feeds and then returns. Average length of these breaks is an hour.

The female, upon leaving the nest, may call, and this may help her locate her mate. She usually joins him while off the nest. He returns with her to the nest, leaving her when they are a short distance away. Neither makes a sound upon returning. The male does no incubating.

As of about the fourth week of incubation, the female and male no longer get together during her rest periods. He may leave the area to join other males and undergo his partial molt.

If the first clutch is destroyed, up to two replacement clutches may be laid.

Nestling Phase

The young hatch within a few hours of each other, and the female does not leave the nest until the young are all able to move about. The female does not feed the young. The day after hatching, the young leave the nest.

The female leads her brood from the nest in the morning. She first periodically looks out of the nest hole, possibly checking for any danger. When she finally decides to bring them out she gives a special call, heard only from very close by, which sounds like "kuk

kuk kuk." At once the ducklings start to crawl to the nest hole, and without any hesitation they jump out with their wings held out, calling softly.

As the young fall from the nest hole they may bounce on the ground a little, and they may, occasionally, be momentarily stunned, but they are rarely hurt. The female and young immediately head for water and feeding areas. This may entail going overland for a distance of over a mile, and it is one of the most dangerous times for the ducklings.

Fledgling Phase

Ducklings and female move to protected spots with readily available food, and may join other families of wood ducks there. The female stays close to the young for the first two weeks; then she may begin to leave them in the early morning and afternoon to feed on her own. The young are generally very independent and, after ten days, may go out on their own.

After about two weeks, several broods may join together and be attended by only one female. Five weeks after hatching, the female leaves the ducklings and goes off to molt with other females. In broods that hatch later, the female may leave the young earlier, for her molt to alternate plumage is imminent. Some females in warmer climates may produce a second brood if their first early one is successful.

The young can fly when eight to nine weeks old. They join together with other juveniles and may form roosts at night. *See* Flock Behavior

Plumage

DISTINGUISHING THE SEXES In all plumages the male can be distinguished from the female by the red at the base of his bill, his red iris, and the two bands of white projecting from the white chin patch. The female has a gray bill, a brown iris, and one band projecting from her chin patch.

Bob Hines

American Woodcock
Scolopax minor

WOODCOCKS ARE BEST KNOWN FOR THE SPECTACULAR FLIGHT-DISPLAYS OF the males over the breeding grounds. These occur at dusk and again at dawn, with the birds circling up to hundreds of feet high and then descending in a zigzag flight. Once they have landed, their distinctive "peent" calls penetrate the silence of the evening or morning hours.

This is the most conspicuous part of woodcock behavior, because of the dramatic flight and calls, but it and other behavior that also occurs in these crepuscular hours are still only partially understood. The purpose of the Flight-displays, how females choose mates, how male and female form pairs, and how females interact with other females are still to be discovered.

Other stages of woodcock life that you may on occasion observe include the time the female spends with her young in summer, feeding in damp, muddy areas. If the family group is disturbed, the female flies up in a fluttering flight that makes her appear injured. This has the effect of drawing any predator away from the chicks. If you look carefully back to where she flushed you might get a glimpse of the young woodcocks. They are extremely cute.

You may also see woodcocks gather into fairly large flocks on summer evenings and during the evenings on their wintering grounds.

BEHAVIOR CALENDAR

	TERRITORY	COURTSHIP	NEST-BUILDING	BREEDING	PLUMAGE	SEASONAL MOVEMENT	FLOCK BEHAVIOR
JANUARY							■
FEBRUARY						■	
MARCH	■	■	■	■		■	
APRIL	■	■	■	■			
MAY	■	■		■			
JUNE	■			■	■		■
JULY	■				■		■
AUGUST					■		■
SEPTEMBER					■		■
OCTOBER						■	
NOVEMBER						■	
DECEMBER							■

DISPLAY GUIDE

Visual Displays

Flight-Display

Male   *Sp Su F W*

The bird lifts off and flies in wide circles. After rising about fifty feet off the ground, its wings start to make the Wing-twitter as it flies higher and in smaller circles. After the bird reaches a height of two to three hundred feet, the twittering stops; the bird gives Flight-song while starting a zigzag descent to the earth. The final portion of the descent is quiet.

CALL: Wing-twitter on ascent; Flight-song on descent

CONTEXT: Occurs mostly in dusk or dawn hours, over breeding grounds. *See* Courtship, Territory

Wing-Raise

Male   *Sp*

As the male approaches the female with a stiff walk or several short runs, he spreads his wings and holds them up over his back.

CALL: None

CONTEXT: Done by the male when approaching a female that he will attempt to copulate with. *See* Courtship

Distraction-Display

Female   *Su*

Bird flushes up and flies a short distance with labored wingbeats and legs dangling down. May be repeated several times.

CALL: None or a sharp cry

CONTEXT: Done by the female when there is a

potential predator near the brood. Possibly serves to lead the predator away from the young. *See* Fledgling Phase

Tail-Fan
Male or Female   *Sp Su F W*
Bird lifts tail vertically and fans it, displaying white underneath.
CALL: None
CONTEXT: Done by a bird when it is alarmed

Bobbing
Male or Female   *Sp Su F W*
The bird repeatedly bobs its rear end and body while keeping its head and bill still. Often done while walking.
CALL: None
CONTEXT: Done by the female when returning to the nest after being flushed and by lone birds when feeding in the open during the day. May be a complex behavior associated with times when the bird cannot see a predator but is wary of one, indicating to the predator that it is alert and knows the predator's whereabouts.

Auditory Displays

Flight-Song
Male   *Sp Su F W*
A series of chirps given for several seconds of the descent from a Flight-display and stopping suddenly as the bird gets close to the ground.
CONTEXT: Occurs on the descent from the

Flight-display and may function as an advertisement of territory ownership. May also function in mate attraction. *See* Territory, Courtship

Wing-Twitter

Male   *Sp Su F W*

A light twittering sound caused by the movement of the outer three primaries of the wings. Occurs for several seconds during the ascent portion of the Flight-display.

CONTEXT: Occurs in Flight-display and, like the Flight-song, may advertise territory ownership or serve to attract mates. *See* Territory, Courtship

Peent-Call

Male or Female   *Sp Su F W*

A buzzy, nasal rendition of the written sound "peent." Preceded by the soft Tuko-call. Given at intervals of about ten to twenty seconds.

CONTEXT: Given from the ground in and around male territories. Given mostly by male before or after the Flight-display. Occasionally given by the female during interactions with the male. *See* Territory, Courtship

Kakak-Call

Male or Female   *Sp Su F W*

A rapid series of harsh notes often running together into an almost continuous sound.

CONTEXT: Given during aggressive interactions, such as during flight chases; usually occurs on the breeding territory. *See* Territory

Tuko-Call

Male or Female   *Sp Su F W*

A two-part gurgling note, heard only within ten to fifteen yards of the caller. May occur in a series, singly, or just before the Peent-call. CONTEXT: Occurs on the breeding ground and seems to be associated with interactions between male and female. Given its softness, it is probably a short-distance communication. *See* Courtship

There are several other woodcock distress and warning calls, but they are seldom heard.

BEHAVIOR DESCRIPTIONS

Territory

Type: Mating
Size: A few hundred square feet to several acres
Main behavior: Flight-display, chases, Peent-call
Duration of defense: From arrival of males on breeding ground into egg-laying phase

Soon after arriving on their breeding grounds, male woodcocks begin to do Flight-displays in the early evening and again in the morning. These are believed to advertise the male's presence on the breeding ground to both females and males. The display tells other males that this area is occupied and they should stay away. When other males arrive in the immediate area of the display, they may be chased away by the displaying bird; rarely, fights may occur between males competing over a given spot. During these aggressive interactions the Kakak-call may be heard.

The Flight-displays in the evening occur over a period of thirty to forty minutes. Those at dawn may last fifty to sixty minutes. On nights when there is a full moon, the birds may continue to display to some degree throughout the night. High winds, heavy rains, or temperatures below freezing usually inhibit Flight-displays.

A single male may defend several spots in a field, or defend one and then move to another, or even shift from field to field. When a male leaves a given territory, he is soon replaced by another male, suggesting that there is a floater population of males looking for the opportunity to occupy abandoned territories.

What area a given male actually defends is unclear. He usually lands fairly near the spot from which he took off. His landing sites may outline the territory. If this is the case, then the territories can vary in size from a hundred square feet to several acres. Clearly, several males can occupy the same field if it is large enough.

A Display-flight area must have open ground and usually consists of a field with some scattered shrubs. During the day males retreat to cover near these display grounds.

Display-flights continue well into the summer, long after the females have finished mating and are involved with raising broods. These later flights may be those of juvenile males rather than adults. The flights may also start early in the year and sometimes occur even while males are on migration in the spring. Some flights also occur in fall.

Courtship

Main behavior: Flight-displays
Duration: From arrival of female into egg-laying

The mechanisms of courtship in woodcocks are still unclear. It is known that males do Flight-displays on territories and that after landing they may interact with females and copulate with them in those areas. The males continue to display after mating and are believed to be promiscuous. But how females choose males and whether females are also promiscuous is not known.

There does not seem to be any lasting bond between mates,

since after mating the female seems to go off, build the nest, and raise her brood on her own.

Interactions between male and female take place when a female comes to the area where the displaying male lands and both may give the Tuko-call or the Peent-call. The male does the Wing-raise display as he walks stiffly toward the female. She crouches down and he steps on her back and they mate. After this, she or he may give the Peent-call and then she usually leaves the area. The male may resume display flights.

Nest-Building

Placement: On the ground in fields, woods, or brush
Size: 4–5 inches in diameter
Materials: A few twigs or grasses

After mating with the male, the female moves one hundred to two hundred yards away and builds a nest. This is a very simple affair consisting of just a little scrape in the ground with a few twigs or grasses arranged loosely around it. It is generally not under dense cover but out in the open. Even so, the color of the eggs and the incubating female make it extremely hard to locate.

Breeding

Eggs: 4. Buff with brown splotches
Incubation: 20–21 days, by female only
Nestling phase: 1 day
Fledgling phase: 6–8 weeks
Broods: 1

Egg-Laying and Incubation

Eggs are laid one per day until the clutch is complete. The female does all of the incubation, which starts after the last egg is laid. She depends a great deal on her color for protection, so much so that she will not flush off the nest until practically touched.

The female has been observed to leave the nest at dusk to feed, and she may leave at other times as well. If flushed off the nest she may do the Distraction-display when flying off. She returns gradually and slowly, possibly doing the Bobbing display as she does so.

If disturbed early in the incubation phase, she may abandon her nest and eggs; if disturbed during the later stages of incubation, she is more likely to return to the nest.

Nestling Phase

The young hatch over a period of about twenty-four hours and stay in the nest, brooded by the female, as their feathers dry. Soon after that they leave the nest and move about with the female.

Fledgling Phase

After a day or two, the young can feed in the same manner as the adults by probing the soil with their bills. Before that they live off the stored food reserves with which they were born. The female and young usually remain in damp areas where they can easily probe the soil for earthworms and look for insects.

During the first week, the female will brood the young a great deal, especially when it is rainy or cold. At about a week old the young begin to develop primary wing feathers. At about three weeks they can fly short distances.

After four weeks, the young can fly well and are about the size of the adults. They stay with the female for six to eight weeks and are then on their own. During the later weeks with the female they wander more widely than before.

When the family is disturbed by a potential predator, the female will flush up and do the Distraction-display. If the predator follows her, she may continue the display for over a hundred yards and then secretively circle back to her young.

Plumage

DISTINGUISHING THE SEXES There are no color differences between the plumage of male and female woodcocks. The female is substantially larger than the male, but there is no clear way to distinguish the sexes when they are apart. There are a few ways to tell male from female by their behavior. The males are the only ones to do Flight-displays. The females are the only ones to incubate and to care for the young.

DISTINGUISHING JUVENILES FROM ADULTS There are no easy ways to distinguish juveniles from adults without close examination.

MOLTS Adult woodcocks undergo one complete molt per year and this occurs in summer, starting in June and continuing until early September. Interestingly, the last feathers to be shed are the outer three primaries, and these are the ones responsible for the Wing-twitter in the Flight-displays.

Seasonal Movement

Woodcocks generally migrate starting at dusk and continuing on into the night in flocks of from a few to over fifty birds. The flocks usually fly at low altitude and may be loosely or tightly bunched.

The fall migration seems to be greatly affected by the weather. The first hard frosts often seem to make the birds head south, possibly because the ground becomes harder to probe and they can no longer feed in the northern areas. They may move in front of a cold wave as well. Most fall migration occurs in October in the North, with the birds arriving in the South from October to December. On their way south the birds often stop over at certain spots for a day or two to feed.

Woodcocks winter in Gulf Coast and southeastern coastal states. By far the largest concentrations winter in the central portion of Louisiana. Individual birds are known to return each winter to the same spot, just as they return to the same breeding ground.

Woodcocks migrate north as early as the beginning of February. The earlier migrations occur when the weather is warmer. As the birds head north they seem to move into succeeding areas where at least some of the ground is thawed so that they can feed. In such a way, they spend about four to six weeks moving to their breeding grounds. They reach the most northern states by late March and early April, in some cases when there are still patches of snow on the ground. Males and females seem to arrive on the breeding grounds at about the same time.

Flock Behavior

In summer, after they become independent from their mothers, young woodcocks tend to gather in the evening at certain fields. These fields are often the ones used as display grounds in the spring. The young fly into the field soon after sunset and remain there for up to a half hour on clear nights and several hours on damp nights with no moon. There may be forty to fifty birds in the same field.

Large flocks of up to several hundred woodcocks assemble in winter in the South at certain feeding spots. Birds may fly for several miles to reach one of these spots, which are usually damp areas with cover. The birds stay all night in contrast to the short times reported for summer flocking.

Bob Hines

Common Tern
Sterna hirundo

IN MANY AREAS OF THEIR BREEDING RANGE, COMMON TERNS HAVE decreased significantly in number over the past sixty years. For example, in Massachusetts the number of nesting tern pairs dropped from an estimated thirty to forty thousand in the 1920s to about seven thousand in the 1970s.

Many studies have been done to try to determine why the common tern population is decreasing. But rather than there being one answer that fits all cases, there seem to be a number of factors causing the decline.

One factor seems to be the rising populations of herring and ring-billed gulls that nest in areas similar to those where terns nest. Gull populations have risen continuously over the last century, possibly due in part to their ability to feed at dumps, which over the years have become larger and more numerous. The gulls are larger birds and dominant over the terns when competing for nesting sites. Gulls also initiate their breeding cycle earlier in the year than terns; thus they are already nesting when the terns arrive at the potential colony.

Terns tend to nest on small protected islands along the seacoast or on large lakes. There is increasing disturbance of these spots by humans as they seek recreation in boats or build near these locations. These activities can scare the birds off their nests and, possibly, lead to less successful reproduction. Because of this, you should try to observe terneries from a distance, and follow these precautions: never enter a tern colony; never get so near the birds that they dive at you; and do not get so close to the colony during the day or night that you cause them to fly up. This makes for far better watching anyway, since you can then see the birds' normal behavior patterns, rather than just their alarm behavior.

Finally, and again due in part to rising gull populations, there may now be more predation on the eggs and chicks of terns by gulls and other large birds. When terns are scared up, gulls can go in and eat the eggs. Great horned owls may also hunt over these islands at night, scaring the adults and possibly making them roost away from the colony at night, leaving it very vulnerable.

There are efforts underway now to afford greater protection to the terns by protecting the islands from human disturbance and by attempting to limit gull interference at terneries.

With a greater awareness of the tern's needs and behavior, we can more effectively judge what factors are important to its continued survival. The common tern is such a graceful and exciting bird to watch that we should do all we can to preserve it.

BEHAVIOR CALENDAR

	TERRITORY	COURTSHIP	NEST-BUILDING	BREEDING	PLUMAGE	SEASONAL MOVEMENT	FLOCK BEHAVIOR
JANUARY					▓		
FEBRUARY					▓		
MARCH					▓		
APRIL					▓	▓	
MAY	▓	▓	▓		▓	▓	
JUNE	▓	▓		▓			▓
JULY	▓			▓			▓
AUGUST					▓	▓	
SEPTEMBER					▓	▓	▓
OCTOBER					▓		
NOVEMBER					▓		
DECEMBER					▓		

DISPLAY GUIDE

Visual Displays

Head-Down
Male or Female   *Sp Su*

The bird lowers head and points bill down. In more intense versions the tail is raised.

CALL: None

CONTEXT: Given mostly by the male. In territorial encounters the bird faces the aggressor; in courtship interactions, the bird stands sideways to mate. *See* Territory, Courtship

Head-Up
Male or Female   *Sp Su*

The bird raises its head, points its bill up, and turns its black cap away from the other bird. In intense versions the tail is also pointed up.

CALL: None

CONTEXT: An appeasement display given between mates, often as they come together on the territory; also given during territorial encounters. *See* Territory, Courtship

Precopulatory-Display
Male or Female   *Sp*

The male, often while doing the Head-down display, walks in a semicircle around the female. He may tilt the top of his head away from the female and even lean away from her during his walking.

CALL: None

CONTEXT: Done between paired male and female; display shows male is ready for copulation. Always done before copulation, but does not always lead to copulation. A late courtship display. *See* Courtship

Aerial-Head-Down
Male   *Sp Su*

During flight, wings are held up high over back and wingbeats are quick, shallow, and widely spaced. Head is pointed down and fish may be carried in bill. May be done alone but usually done with another bird.

CALL: None or Keirr-call

CONTEXT: Given by male during courtship flights. *See* Courtship

Aerial-Head-Up
Female (or male)   *Sp Su*

During flight, wings are held arched down and bill is pointed straight in front. Usually occurs when bird is passing another bird who is in Aerial-head-down.

CALL: None or Ki-ki-call

CONTEXT: Given by female (or intruding male, early in season) during courtship flights. *See* Courtship

Auditory Displays

Kip-Call
Male or Female   *Sp Su F W*

A very short, single, high-pitched note.

CONTEXT: The most common call at undisturbed terneries. Given by the birds as they fly off territory and as they hover over water in search of food. May act as a contact call between birds.

Kierr-Call

Male or Female   *Sp Su*

A short, drawn-out, descending call, often repeated. Less harsh and shorter than the Keeearr-call. May intergrade to "keeri-keeri-keeri," ascending at the end.

CONTEXT: Usually given by bird as it approaches mate or young with food. May be started soon after bird makes catch and continued all the way back to the mate or young.

Keeearr-Call

Male or Female   *Sp Su F W*

A drawn out, shrill cry lasting about one second.

CONTEXT: An alarm call, given in situations of danger. In mild excitement it is repeated slowly with relatively long intervals between calls. With increasing excitement, calls become shorter and more rapidly repeated. Often given when another bird starts to trespass on territory or during human approach to the ternery.

Kek-Kek-Call

Male or Female   *Sp Su*

A short, low-pitched, sharp sound of "kek" rapidly repeated in a long series: "kek-kek-kek-kek-kek-kek."

CONTEXT: Call given in situations of extreme danger or during aggressive interaction with another tern, such as a fight with or aerial attack on a territorial intruder. Most often given by males. *See* Territory

Ki-Ki-Call

Male or Female   *Sp Su*

A high-pitched, shrill, rapidly repeated cry:
"ki-ki-ki-ki-ki-ki"

CONTEXT: Given by female or, occasionally,
male while begging for fish from another
bird. *See* Courtship

BEHAVIOR DESCRIPTIONS

Territory

Common terns may defend two types of territories during the
breeding season. One is the nesting territory and the other is the
feeding territory.

Nesting Territory
Type: Nesting
Size: Averages about 4 feet in diameter
Main behavior: Keeearr-call, Head-down, Head-up
*Duration: From a few weeks after arrival at ternery until midway
through fledgling stage.*

Common terns, when they first arrive on the breeding grounds,
do not settle at once in the ternery but, instead, roost at night on
nearby beaches and rocks and feed in the area during the day. The
birds may arrive in the area gradually over a period of a month and
during this same period slowly filter into the ternery. The later a
bird arrives in the season the greater its tendency to go directly to
the ternery to start its breeding behavior. Older terns are generally
the first to arrive and also the first to settle.

In the first week after their arrival, the birds may begin to make
morning and evening flights over the ternery. Later, the birds will
begin to land on the ternery. At first they may land at several
different locations, but gradually they restrict their landing to a

specific spot and defend it against other terns. Among paired birds both male and female may defend the territory. Among unpaired birds, it is the male that initiates territory defense. Once a territory is claimed, the birds spend more time on the ground defending it.

Terns have a strong tendency to occupy territories from the previous year. Young males, who arrive later, are generally forced to occupy less-favorable territories. Territories are occasionally large, up to six feet or more in diameter, if there is little population pressure. In crowded situations, territories may be as small as a foot and a half in diameter. In general, the smaller the territories, the more aggression occurs.

A typical territorial encounter involves a male territory holder and another bird that begins to intrude. The territory holder at first gives slow Keeearr-calls or Kierr-calls. If the intruder continues, the holder gives calls more rapidly and raises its wings. At the same time, both usually do the Head-down, facing each other. If this does not stop the intruder, then the territory holder attacks him, giving the Kek-kek-call, and the two grab at each other's neck and head. These fights can be severe, but generally one bird soon gives up, assumes the Head-up display, and leaves.

Males are generally more aggressive than females, but the female may join the male briefly in fighting off an intruder. Joint defense between neighboring pairs may involve synchronized displays in which both pairs alternately do Head-up then Head-down displays. This performance may be accompanied by the Kierr-call. Occasionally, an incubating female tern will leave her nest and attack an intruder on a neighboring territory. Following that, she may even briefly sit over the eggs in that territory before returning to her own nest. It is not known why she does this.

A tern may also defend its territory from the air by diving down upon an intruder while giving the Kek-kek-call. Or neighboring males may fly up, facing each other, with their tails spread and their wings rapidly beating. This may be accompanied by the Kip-call and Kek-kek-call. The two spiral around each other high into the sky, then at a certain point they cease their upward flight and fly about, seemingly unconcerned with each other, but still using rapid wingbeats. The birds return to ground level and then the performance may be repeated. Sometimes this flutter flight is done by a lone bird. Its significance in this case is unknown.

Besides the nesting territory, an individual bird may claim a perch near a fishing area and at times defend this from others, directing the Head-down display at intruders. This same behavior can be seen among gulls on rooftops and dock pilings.

Defense of territory begins to lessen about two weeks after the young have hatched, and soon after that the territory is not defended at all.

Feeding Territory
Type: Feeding
Size: 100 – 300 yards of shoreline
Main behavior: Chases
Duration of defense: Usually just early summer

In addition to nesting territories, terns in some colonies also have feeding territories. A typical feeding territory runs along one hundred to three hundred yards of shoreline, and includes the shallow water and some of the land. Within the territory, the birds may have favorite perches and fishing spots used day after day. Feeding territories are used most in May and June but they may be defended throughout the breeding season. They may be up to five or more miles from the ternery.

The terns use these territories for about a third of the day, the rest of the time being spent on their breeding territory, feeding or resting. For a week or two prior to egg-laying both the female and male may use the feeding territory — the female resting on shore and being fed by the male.

Defense of the feeding territory is primarily by the male and involves the male flying low out over the water toward an intruder in a fast, direct flight and pursuing the intruder for a hundred yards or more out of the territory.

Courtship

Main behavior: Aerial-head-down, Aerial-head-up, Head-down, Head-up, mate-feeding
Duration: From shortly after arrival at ternery until incubation

Terns do not seem to breed until they are three to four years old. Before that time they do not seem to return to the breeding ground. Where they stay is not known. Occasionally, young terns may return to the ternery but will be unsuccessful at breeding. During the night and, sometimes, during the day, these birds tend to roost in groups outside the ternery. They may be joined by adults that are single or who were also unsuccessful breeders.

Paired terns continue to breed together from year to year. Whether pairs remain together through the winter is not known, but when the birds arrive near the ternery and roost together, many are clearly already paired, as reflected by their standing close together.

In about the second week after the terns arrive in the area of the ternery you may begin to see courtship. Terns do courtship displays in the air and on the ground. Aerial courtship involves the male flying along, often with a fish, followed closely by the female, who is either above or to the side of him. At some point she begins to pass him. During this pass he does the Aerial-head-down display, with his head turned away from her and his wings high over his back, while she does the Aerial-head-up display with head straight forward and wings held low.

Having passed the male, the female starts a fast, steep downward glide during which she tilts from side to side. Each may pass the other several times as they glide, and whenever they do so, they do their respective Aerial-head-down and Aerial-head-up displays. Near the ground the flight levels off. The two may repeat this performance or stop. Occasionally, a third bird joins in these flights.

In ground courtship the male walks in a semicircle in front of the female. He usually does the Head-down display with the top of his head turned away from her. He may carry a fish in his bill. This is believed to signal his intention to come closer to the female and copulate. If she is unreceptive to the male, she may move away from him. This may stimulate further display by the male.

If she is receptive to him she may do the Head-up display. At this point he may also do the Head-up and move closer. If she is ready to copulate, she crouches down, he steps on her back, and they copulate. He may stand on her back for several minutes, and they may copulate several times. During copulation the female points her bill up towards that of the male. After the male gets off the two may do the Head-up display and then preen.

During courtship displays on the territory, either or both members of the pair may do scraping, which entails their bending forward and kicking back with their feet. This is how they build their nest, but scraping is also done at other times, possibly signifying ownership of the territory and intent to breed. *See* Nest-building

In the week prior to actual copulation, you may see one bird flutter onto the back of another, just stand there for up to a minute, and then get off. This action occurs primarily on the territory, while actual copulation may occur anyplace the birds frequent.

Courtship-feeding or mate-feeding is another common element in tern life. During mate-feeding the female assumes a hunched posture and gives the Ki-ki-call; the male then gives her the food. In the early stages of courtship, males may take fish in their bill and fly low over the ternery, possibly to attract the attention of females. Occasionally these flights result in the male's giving the fish to a female or even to another male that may deceptively act like a female to get the fish.

Once pair formation has occurred the two birds may spend more time at a feeding territory (see Territory) where the male mate-feeds the female often. The pair occasionally return to their nesting territory during the day and spend the night there.

This phase may last from five to ten days. Several days prior to egg-laying, the female returns to the nesting territory and stays there while the male brings all food to her. This continues up until the laying of the first egg. From this point on, both birds share in incubation, and mate-feeding by the male decreases rapidly.

Nest-Building

Placement: On the ground, in the open or near vegetation
Size: Varies from individual to individual, but usually a few inches in diameter
Materials: Just a scrape in the ground, occasionally encircled by pebbles, straw, sticks, and other objects collected in the nest's vicinity

Common terns generally nest in colonies of from a few to almost two thousand pairs. A typical colony has about two hundred pairs. The birds nest next to water, either on the coast or along inland rivers and lakes. Inland colonies tend to be smaller than those on the coast.

Nest-building for common terns primarily involves scraping a shallow depression in the earth. To do this the bird bends forward and scrapes back with its feet. This scraping activity starts as soon as the birds begin to land in the ternery. Many scrapes may be started by the male; this seems to stimulate the female to enlarge them and, perhaps, add bits of debris around their edge.

Scraping and nest construction continue like this at a low level until a few days before egg-laying, when they occur more often and with greater intensity. Once she is ready to start laying the eggs, the female chooses one of the scrapes in the territory. The nests are often so shallow that they barely keep the eggs from rolling out.

Breeding

Eggs: Usually 3, occasionally 2 or 4. Buff with dark brown spots in a wreath or concentrated at one end
Incubation: Usually 21–22 days, but may last up to 31 days; by male and female
Nestling phase: About a week
Fledgling phase: 3–4 weeks or longer
Broods: 1

Egg-Laying and Incubation

The average clutch contains three eggs. The intervals between layings in a clutch increase with each egg; generally one day between the first two eggs and two days between the second and third eggs. Incubation starts after the laying of the first egg, which, of course, results in the young hatching at different times. Incubation usually lasts from twenty-one to twenty-two days, but may last up to thirty-one days depending on the weather and how much the birds are disturbed during the incubation phase.

Both male and female share in the incubating but the female generally does more than the male. The member of the pair that is not incubating may feed, get food for the incubating mate, or rest on common ground overlooking the nest. When exchanging places at the nest, the nonincubating bird may simply initiate the switch by raising its wings, which makes the other bird fly off; or it may offer the incubating bird a fish to get it off the nest; it may also preen the incubating bird slightly. The incubating bird, before leaving the nest site, usually tosses debris over its shoulder and gives the Kip-call; then it takes off.

Incubation periods for each bird range from a few minutes to a few hours. On very hot days, the birds stand over the eggs to keep them from overheating.

Terns and gulls are different from most other birds in that they have three incubation patches. They also most commonly have three eggs. A recent study showed that birds with three eggs incubated more constantly and with less disturbance than those on one- or two-egg clutches. It was also found that a significant number of terns with fewer than three eggs place stones of about egg size in their nest, perhaps simulating the missing egg and promoting better incubation.

Nestling Phase

The young hatch about one and a half days apart, usually during the daylight hours. At the time of hatching the shell remains are taken away from the nest by the parents. After hatching the young may crawl out of the nest. During the first five to seven days, they generally crouch together and are brooded, mostly by the female. Since the young are precocial and covered with down when they hatch, there is no specific nestling stage and they may be brooded anywhere on the territory.

The male does all of the fishing at this time and supplies food for himself, the female, and the young. As of about a week after hatching, both parents may go out to catch fish for the young.

After the first week, the chicks may wander about the territory, hide next to grasses and stones when resting, and only come out when the parents arrive with food. A parent brings back a fish in its bill and presents it directly to the young bird, who eats it whole; if the fish is laid on the ground the young will not eat it. Young birds peck at the parent's bill when they want to be fed. The parents' tendency is to feed the oldest first since they beg most strongly.

Fledgling Stage

When the young are not being fed they often peck at random objects in their immediate environment. They also may wander off the territory since they have no knowledge of its boundaries. In this case the neighboring territory holder may attack them if they move actively about, but if they crouch down the adult will not be aggressive and may even brood them for a moment and then leave them alone. The young birds may scrape out an area in the sun in which to rest, perhaps for added warmth. Young from different parents may gather together as they get older in places away from their territories. When the adult approaches with food, the young will return to the territory to be fed.

After about a week you can see juvenal feathers on the young. At three weeks of age the young flap their wings and try to fly. By four weeks they have gained all of their juvenal feathers and can fly, but not very strongly. If they are scared up at this time they may fly out over the water and, because of winds, not be able to make it back, so be cautious about disturbing them at this time.

Even when the young can fly, they tend to sit and wait for the adults to bring food, begging noisily as the adult approaches. After about a week of this, the young bird begins to fly up to the adult as it comes near with food, but it is still fed when they both land. After another week the young bird begins to follow the adult as it searches for food.

The young do not fly that much even once they are able to, but continue to grow and gain weight. They may weigh even more than adults prior to migration, and this has obvious survival value for them. In some cases the young are still fed by their parents after migration is underway, or, in the young of late-nesting pairs, even on the wintering ground.

After the fledgling phase the young may join into small flocks or remain in family groups.

Plumage

DISTINGUISHING THE SEXES There is no easy way to distinguish male from female by plumage or size. Even behavioral clues to distinguishing the sexes are scarce. The male is the first to establish the territory and also, generally, the bird carrying a fish in courtship flights.

DISTINGUISHING JUVENILES FROM ADULTS Common terns do not develop full adult plumage until they are three years old. The main feature distinguishing the juvenile from the breeding adult is an incomplete black cap with white still remaining on the forehead, which is unlike the adult's plumage, where the black cap reaches completely to the bill. In winter, adults and immatures look similar.

MOLTS Adults (three-to-four-year-old birds) have two molts per year — a complete molt and a partial molt. These molts overlap. The complete molt may begin in August and September, but is arrested during migration and then continued on the wintering ground. This molt is completed, gradually, by early March. A partial molt of body, tail, and some wing feathers overlaps with this, starting as early as December and finishing around April.

Molts of immatures occur at different times and in different sequences from those of adults.

Seasonal Movement

Common terns migrate north from their wintering grounds in late April and early May. They migrate in small flocks but may join into larger temporary flocks as they stop to feed at areas rich in food. In both spring and late-summer migration, the birds often fly high and may fly during day or night.

Southward migration after breeding occurs from mid-August into the first weeks of September. Again, the birds generally fly in small flocks although large flocks of over a thousand birds have been reported. In some cases, when young are still being fed by their parents, they may begin migration as a family. Common terns winter from Florida and southern California to the southern tip of South America.

All ages of common terns winter together, often in the company of sandwich, roseate, and black terns.

Flock Behavior

As you watch a tern colony, one of the most notable features is the constant activity. Much of this activity is synchronized, with all terns in the colony participating. For example, when one bird gives the Keeearr-call and flies quickly up from the colony, it often stimulates others near it to fly up also, but quietly. This quick "upflight" is the common reaction to a predator. Once the birds are in the air, they all give the Keeearr-call noisily and dart down over the head of the potential predator until it leaves the area. The Keeearr-call may become more intense and change to the Kek-kek-call as the birds dive at the predator.

In other cases, if one bird quickly and quietly flies low off the ternery and out to water, it may stimulate others near it to do likewise. These quiet desertions of the colony are called "dreads," and it takes a minute or so until the colony becomes noisy and settles down again. Dreads are presumably done in response to possible immediate danger.

There are also unexplained variations in the excitement level of a colony — sometimes it may seem very excited and noisy and at other times it may be relatively quiet. Also, the predominant level of excitement may vary from colony to colony. It may be that larger colonies are less susceptible to fear of a predator than small ones — in other words, there may be a feeling of safety in numbers.

Common terns regularly gather as flocks for roosting at night and/or for resting, bathing, and preening near fishing areas or just outside the ternery. These common grounds are usually on open flat areas near water, such as mud flats, flat rocks, or sandbars. They are usually at or above high-tide level. These roosts may even be used by breeding birds during incubation and nestling stages, but usually only when there is nightly predation on the ternery, such as by a great horned owl.

Terns in flocks may do Head-down or Head-up displays while maintaining a certain amount of distance from each other. Paired birds will stand closest together.

quarter mile away. This demands the use of a scope or strong binoculars. We must resist our desire to get the closer photograph, or a more intimate look, in favor of letting bald eagles prosper.

BEHAVIOR CALENDAR

	TERRITORY	COURTSHIP	NEST-BUILDING	BREEDING	PLUMAGE	SEASONAL MOVEMENT	FLOCK BEHAVIOR
JANUARY							■
FEBRUARY						■	■
MARCH	■	■	■	■		■	■
APRIL	■	■	■	■	■		■
MAY	■		■	■	■		■
JUNE	■		■	■	■		■
JULY	■		■	■	■		■
AUGUST	■			■	■		■
SEPTEMBER	■			■	■	■	■
OCTOBER						■	■
NOVEMBER						■	■
DECEMBER							■

DISPLAY GUIDE

Visual Displays

Talon-Drop
Male or Female   *Sp Su F W*
The bird lowers its talons when approaching another eagle in flight from above. Just before they touch, the second eagle rolls over and presents its talons. The two may touch briefly.

CALL: None

CONTEXT: Seen most in winter and spring when groups of eagles are soaring together. *See* Courtship

Auditory Displays

Bald eagles are generally silent and, in particular, do not make a lot of noise around the nest. Although their vocalizations have not been well studied, the following calls seem to be important to their social behavior.

Scream-Call
Male or Female   *Sp Su F W*
A piercing scream that may be repeated seven or eight times, or more. During call the eagle gradually tilts its head up and back.

CONTEXT: Given between paired male and female during nesting. May serve alternately as an alarm call, a long-distance contact note, and/or an aspect of pair-bonding. May be given by both birds at the nest, when one bird is away from the nest, or when intruders such as crows or ospreys fly overhead. May be accompanied by rapid wing movements.

Chitter-Call
Male or Female   *W Sp Su*

A rapid series of chirping sounds, often trailing off at the end.

CONTEXT: Given when flying over nest, when exchanging places on the nest, when initiating any new act around the nest, or after chasing away an intruder.

Nestling-Calls
Male or Female   *W Sp Su*

Screams that, during the nestlings' development, begin to sound more and more like the Scream-call of the adults. These include a variety of calls from a quiet "yeep" to a rising squeal to a shrill scream.

CONTEXT: Given when the young see the adults flying toward the nest with food, or when the young are being fed or fighting over food on the nest.

Territory

Type: Nesting, mating, feeding
Size: Area around the nest and part of water area for feeding
Main behavior: Aerial chases and dives
Duration of defense: Throughout the time they remain on the breeding ground

Bald eagles are territorial, defending the area around the nest site, which often includes part of a coast, lake, or river shoreline where the birds can feed. Once a pair is established on a territory, they are very reluctant to move elsewhere to breed. Even when all suitable nest trees have been cut or have fallen in storms, some mated pairs have remained for a season or more on their territories without breeding, rather than moving to a new territory.

The birds return to the territory for years, and over that time, one or several nests may be built on it. Eagles, whenever possible, stay on their territory throughout the year, only leaving when food becomes too hard to find or weather conditions too severe.

A good territory has a suitable nest tree, lots of good perches where the birds can sit and see all of the surrounding area, good perches at strategic points around the perimeter of the territory, and a good feeding area.

The nest site becomes a focal point for the pair throughout the year, where they may perch together and bring food to eat.

Territorial interactions between eagles involve the territorial bird, male or female, diving down on the intruder while it is in midair, and the intruder rolling over on its back at the last moment to present its talons to the charging bird. Either one or both of the pair may attack the intruder. Intruders are often subadult bald eagles.

The territory may extend up to a half mile in each direction from the nest. Nests in crowded areas with good fishing may be as close as 150 yards apart, but the average distance between nests is one to two miles.

In areas where there are lots of eagles, other eagles flying through a pair's territory at two to three hundred feet from the nest are generally not bothered by the owners. The owners may just give a Scream-call. In one case, a subadult eagle remained in the area of the nest, perching thirty to forty feet, or more, from the nest during the breeding season, and was not bothered by the adults. This young bird may have been from a previous year's brood by the pair. In other cases, as noted, subadult eagles are chased off a mated pair's territory.

The most aggression seems to occur around birds nesting for the first time. Perhaps, in successive seasons, neighboring eagles get to know each other; they then have fewer aggressive interactions, possibly because they know and respect each other's territorial borders.

Other birds that come too near the nest may also be chased or attacked by the eagle. This is especially true during the periods of egg-laying, incubation, and nestling. At these times an eagle will fly out after a crow, gull, or other hawk until it has left the area and, again, possibly even attack it. During other phases of breeding and at other times of the year, these smaller birds may dive at and harass the eagle, and the eagle may not reciprocate in any way except to move and try to avoid them.

Courtship

Main behavior: Soaring together, chases, dives
Duration: Winter through spring

There are not many obvious signs of bald eagle courtship. Pairs seem to arrive at the breeding ground together and remain together for as long as they both live; an eagle surviving the loss of its mate gets another one. How this occurs is not known, but in one case, a female, after losing her mate, left the territory and returned about three months later with a new mate. This one female may have had up to four replacement mates in her lifetime.

When eagles are sexually mature is not known for sure. Some researchers believe it is as early as their first spring. A bird in full adult plumage may be mated to another that still bears signs of subadult plumage, yet successfully raise young.

In late winter and early spring, when eagles are congregated at feeding areas or migrating, you may see many eagles all soaring together, sometimes over an area of several square miles. The soaring is particularly common in late morning and continues through the afternoon, possibly because thermals are strong enough to lift the birds at these times of day. You may see chases between multiple pairs of birds, and chases where one bird dives at another, with the attacked bird rolling over at the last second and extending its talons at the diver. There is rarely actual contact between the birds. These aerobatic maneuvers may occur to some extent at any time of year but, again, they are especially common in spring and late winter, leading many researchers to believe they are associated with courtship. These chases often involve subadult bald eagles or subadult and adult birds, but do not occur as often between two adult eagles.

Chases usually occur in a slight glide and the chased bird may swerve to either side or up and down to elude the chaser, often in dives followed by climbs. Chases may last up to eight minutes. Chaser and chased may change places, sometimes after a dive. Talon-drop occurs as one bird dives on the other. Occasionally talons are locked and the hawks fall through the air, whirling. Also, occasionally on the winter ground or summer ground eagles may pass twigs in midair. In some cases, two birds that have been chasing one another may come down and perch near each other after the chase.

Near the nest site, or on perches at the nest, you may see some close interactions between a mated pair. First of all, the birds will perch next to each other. If they are very close, one bird may "bill" the other, which entails lightly pecking at the other's bill. One bird may preen the other by running its bill through feathers on the other bird's head, neck, back, or breast regions. Occasionally, the two birds may lie right next to each other in the nest, and billing or preening may occur at these times also. All of these activities seem to occur most often in the early morning or late afternoon.

Copulation takes place on or near the nest. The male steps onto the back of the female with talons closed and may flap his wings, possibly for balance. He may call. She raises her tail while he lowers his to make contact. He only stays on her for a few seconds, then gets off. The two may then perch together for up to half an hour, and one or both may preen and/or arrange material on the nest.

Several cases of three bald eagles remaining in the area of one nest have been reported. And in some cases the nest then had four eggs, the norm being two. Occasional polygamy may occur in bald eagles, with two females laying in the same nest. It is thought that this may occur particularly in food-rich areas.

Eagle pairs seem to stay together throughout the year.

Nest-Building

Placement: In crotches of large trees or on cliffs
Size: 6 feet in diameter and 4–10 feet deep
Materials: Base of sticks and branches, lined with sod, grasses, sea-
weed, and other materials

Although nests may be up to a mile from water, they are generally located near or at the edge of a lake, large river, or seacoast, all
of these being places where the birds feed. Nests are usually built
in the crotches of large trees, but when these are unavailable, the
eagles may build on cliffs. Bald eagles build tremendous nests
needing large and strong supports. Trees must usually be over one
foot in diameter and have large, strong forking branches at the top.
Eagles prefer to nest in the tallest trees in a given area, often ones
over a hundred feet high, which gives them a good view of their
surroundings.

Nest-building may start in September or October in the South,
but, generally, does not start in the North until January to March,
or whenever the pair arrive back on their territory if they had to
leave for better feeding grounds. The pair will return to the same
territory each year and may build on the same nest for over twenty
years (if they live that long), or may have two or three nest sites in
their territory among which they switch from year to year.

Both male and female help build the nest. The nest in the first
year is a mass of large branches, some up to two inches in diameter
and six feet long, although generally about half that long.
Branches are usually collected off the ground, but there are reports
of eagles breaking off dead branches in their talons as they fly by
them. They are placed in the nest with the bill and arranged to
form a circular mass, often cup shaped, up to six feet in diameter
and about four feet deep. The middle of the nest is then filled with
sod, weeds, grasses, seaweed, Spanish moss, and other plant material. In the South, eagles have also been known to carry odd
objects such as light bulbs, trash, and string to the nest.

It is in the center portion of the nest that a small hollow, about a foot in diameter and several inches deep, is made and in which the eagles will lay their eggs. The whole nest can be completed in four days' time.

Eagles returning for a second year often build right on top of the previous year's nest. This may add about a foot or two of material to the top of the nest. Over the years this leads to a considerable accumulation in both size and weight. One record nest was nine and a half feet in diameter at the top and twenty feet high; another, after it had fallen, was estimated to weigh two tons. After a while, most nest trees collapse under the weight, usually during strong winds.

Eagles continue to bring material to the nest throughout the incubation and nestling phases. Like many other hawks, the bald eagle has the habit of bringing sprigs of fresh green leaves from deciduous or evergreen trees. The function of these is not known for sure, but several observers have seen eagles eating this greenery. Also, records of eagle pellets show they sometimes contain large amounts of vegetable matter. Thus the sprigs may function in the bird's diet or digestion in some way.

Breeding

Eggs: 2. Dull white
Incubation: 34–36 days, by male and female
Nestling phase: 10–12 weeks
Fledgling phase: 3–4 months
Broods: 1

Egg-Laying and Incubation

Either the male or female bald eagle may be involved in pseudo-incubation before the eggs are laid. This can occur for several days before laying and can be confusing to the observer, since the bird acts as if it is incubating.

The typical clutch contains two eggs. They are laid two to four days apart and, in some cases, possibly a week or more apart. Occasionally, three or four eggs are found in a nest. Some researchers have suggested that this may be the result of polygamy.

Incubation begins soon after the first egg is laid and lasts thirty-four to thirty-six days. Both male and female incubate the eggs, sharing day and night duty. In general the female does slightly more. When they change places, one comes down to the nest, often from a nearby perch. One or both birds may give a soft chittering call. The birds walk carefully when near the eggs, possibly so as not to injure them with their talons. The incubating bird usually probes the eggs with its bill before settling over them, and, after getting settled over the eggs, it usually rakes nest material around its body with its bill. Typical stints at incubation last one to three hours.

During incubation, the bird is hard to see on the nest, as it often hunkers down, but it may peer over the edge of the nest. The eagles may be very quiet when approached during this time and may not give any alarm calls. While one bird is incubating, the other may be either hunting or perched nearby and on the lookout. When the birds leave the eggs for any amount of time, they usually cover them over completely with the vegetation added to the nest so that they are not visible.

Nestling Phase

The young hatch several days apart. For the first two weeks the parents brood the young frequently; after that, brooding is done less often. Brooding often follows feeding and can be done by either male or female. When brooding, the adult may, as when incubating, pull softer nest materials about its body to help keep the young warm. When the adults exchange positions at the nest they may give the Chitter-call. The newly arriving adult often walks carefully about the nest inspecting it or rearranging nest material, which it continues to bring. Then it settles down over the young to brood.

The young are watched almost continuously, with one adult always at or right next to the nest, feeding on prey remains, feeding the young, sheltering them from sun, rain, or cold, or perched and looking in all directions. At some nests the adults will remain silent even when approached.

A parent bird may remain perched at the nest for two to three hours at a stretch. Occasionally, when it is too hot, the adults may perch lower and in the shade. Both young and adults, when overheated, will let their wings droop down and away from the body — probably allowing for greater circulation of air. They may also hold their mouths open.

The typical behavior of a feeding adult is to leave its perch near the nest, taking a short flight over the nest, possibly circling it, and then flying off to get food. Within a half hour or less it will return, land on the nest and feed itself and the young, then go to its perch. Adults may give the Scream-call while at the nest, but not necessarily at the times of arriving and leaving.

Male and female may both feed the young, or one adult may pass food to the other, which then feeds the young. Most feeding trips to the nest are early in the morning or from mid- to late afternoon. Four to eight feeding trips are common in a day.

When the adult arrives at the nest with food, it may tread upon the prey to break it up, then tear off bits of flesh to feed the young, bill to bill. Direct feeding of the eaglets, bill to bill, may continue until the day they leave the nest, and even after, or it may stop up to two weeks before they fledge.

In the later nestling stages, the parent may just drop off the prey at the nest. One of the nestlings may then claim it by crouching over the food with wings spread and head low, screaming. The bird that does not get food may redirect its attention to a stick or other object in the nest, picking it up in its bill, or tossing it about. The adults occasionally bring clumps of grasses and leaves instead of food. The young may peck at this or play with it.

In many cases, the firstborn dominates the second born in competing for food, and thus grows faster. This differential may even lead to the death of the second born. This is a fairly common occurrence among bald eagles, and, in some other eagle species, an event that almost always occurs. It is believed that this is a way to match the size of the brood with the available food resources. If there is not enough, only the oldest lives; if there is enough, then both live.

The physical development of the young can be roughly divided into two stages. The first stage lasts about five to six weeks, and during this time the young are covered with grayish down.

In the second stage, which also lasts five to six weeks, the young develop mature feathers. At the start of the second stage, the young eaglets do a great deal of preening, which has the effect of loosening the down. Downy feathers at this time may litter the edges of the nest, fly off the nest in wind, and fall into the surrounding trees, alerting you to this stage of nestling development.

In the first few weeks, the young are hard to see in the nest since they may flatten down when danger is perceived. At this age they may also crawl around in the nest, using their wings to help them move.

In later weeks, the young begin to perch about the edge of the aerie and are more active. Their activities include flapping their wings, preening, hopping about on the nest, resting either perched or lying down, feeding, playing with objects on the nest, carefully scanning the area around the nest, and treading on food items or parts of the nest. Actions performed by one eaglet often stimulate the same actions in the other eaglet. The eaglets also back up to the edge of the nest and shoot their feces out away from the nest. This may create a ring of whitewash below the nest.

As the time of fledging approaches, the young take practice flights off the nest, hovering into the wind several feet off the nest, possibly circling out a little, and then returning. Finally, they take their first flight, which may vary in length from a short trip to a neighboring tree to a journey of almost a mile.

The nestling phase lasts from ten to twelve weeks.

Fledgling Phase

There are roughly three stages of fledgling life for bald eagles. For the first four to six weeks, they remain attached to the nest site and the parents and can usually be found within a half mile of the nest. They may use the nest as a perch, feeding site, or roosting place. They are brought food by the parents but also begin to forage on their own and may follow the parents on their hunting trips.

In the second stage, which lasts several more weeks, they lose their attachment to the nest and wander, often staying along shorelines for the good hunting to be found there and often moving in the direction of the stronger prevailing winds.

In the final stage, seven to eight weeks after fledging, they may develop a stronger instinct to move. If they are in an area where eagles migrate, they may begin migration. If they are in an area where eagles do not migrate, they may just move to areas where there is an abundance of food and they can feed more easily. During this stage they may sometimes be driven off by the adults.

Plumage

DISTINGUISHING THE SEXES Male and female eagles are similar in plumage. The female is generally larger than the male, but this is a helpful clue only when the two are near each other. There is no obvious behavioral difference between the sexes, since both sexes share responsibilities at the nest fairly equally.

DISTINGUISHING SUBADULTS FROM ADULTS Bald eagles do not fully gain their adult plumage until they are three to four years old, or, possibly, older. During this time, they are called subadults. In their first two years they are mostly dark brown all over with a mottling of white on their underwing coverts and tail. In the next year the head becomes whiter but the tail remains mottled white on brown. In the third year the head and tail are dull white. In the fourth year the head and tail are bright white and the body feathers all dark brown.

MOLTS Bald eagles seem to molt once a year, possibly starting at the beginning of the breeding season and continuing through until the fall.

Seasonal Movement

What is known about bald eagle migration suggests that there are different patterns of migration depending on where the eagles have spent the summer. In general, adult eagles seem to stay as near to their nesting territory as the food availability and weather conditions will allow. This means year-round territory occupation for some eagles, and for many others it means the entire year except for a month or two in midwinter when they move to better feeding areas. Subadult eagles that have not yet established a breeding territory are believed to wander more. With hunting skills less developed than adults', they must look for the easiest food available throughout the year. This means they may roam in both summer and winter. *See* Flock behavior

In northwestern coastal areas, eagles after breeding may move to river areas where there is extensive salmon spawning to take advantage of this source of food. This often means flying north.

Subadult eagles seem to stay there longer than the adults. The adults soon go back to their territories to fish along the coast and defend this feeding area. Subadults must wander further and feed on carrion. Thus the populations of eagles are somewhat segregated by age.

In north-central North America, eagles breed on lakes and along large rivers. In northern parts of central Canada, eagles migrate south in winter when their lakes and rivers freeze and they can no longer get food. They tend to move south along large rivers and stop when there is open water where food can be had. They often gather in large numbers at these spots and roost and feed together. Fall migration is more leisurely than spring migration, for the birds stay at good feeding areas until the food runs out and only then do they move on, often spending one week at one spot and a week at the next. During spring migration they may move directly back to the breeding areas.

Migration in the Northeast, where the birds nest primarily on the coast, is similar to that in other regions in that adult eagles remain in or near their breeding territories as long as food is available and subadult birds tend to move away to other areas where more food is available. Movements can be irregular and may be westward or eastward as well as southward.

In Florida and along the Gulf Coast, eagles breed in November and December and make a reverse migration north in late spring and early summer. They move up along the Atlantic coast or up the Mississippi River valley. In September these birds then migrate south to breed; they are the eagles most often seen at eastern hawk-watch sites in fall, since eagles that breed in the North do not migrate until November or December.

RECOGNITION DURING MIGRATION A huge, dark bird with long and broad wings. Soars continuously, only rarely giving deep flaps with its wings. White head and tail of adults are distinctive; subadults are dark with blotches of white on parts of wings and tail nearest the body. Sometimes referred to as a "flying two-by-four" because of its steady soaring and plank-like appearance.

BEHAVIOR DURING MIGRATION Eagles may be seen as lone birds or in groups as they migrate, depending on the population of eagles in the region. It is typical in the East to see a lone eagle in among a group of broad-winged hawks; the eagle will rise more slowly on the thermals due to its greater weight.

For more information on hawk-watching, see Appendix B, Hawk-Watching.

Flock Behavior

Depending on the location, large aggregations of eagles can be seen at any time of year. In fall and winter, large flocks of adult and immature eagles can be seen at areas where there is adequate food, such as open water, places where there are concentrations of wintering waterfowl, salmon spawning areas, and areas with carrion (sometimes provided by state fish and game departments). In some cases there may be as many as a hundred eagles in a group. At night, the eagles generally use communal roosts that are within a mile of their feeding areas. They leave the roosts in the morning and may start back to them as early as 2 P.M. On days with bad weather they may remain in the roost. Roost sites and feeding areas may change according to the availability of food.

During the summer, there may also be large concentrations of eagles around feeding areas, bathing areas, or possible roosting areas. It is assumed that most of these eagles are nonbreeding birds, but why they congregate together is not known.

Bald Eagle Fall Migration Chart

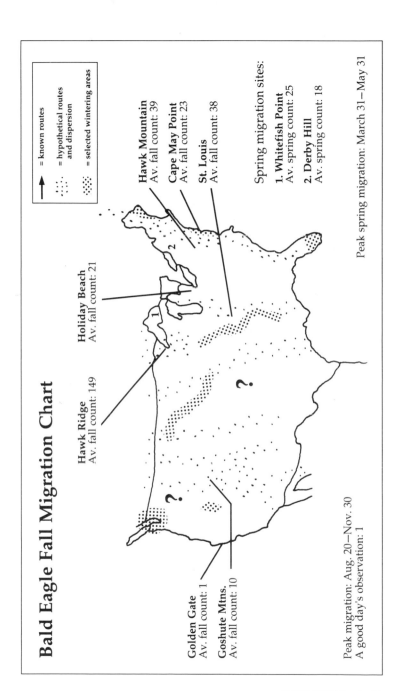

= known routes

= hypothetical routes
and dispersion

= selected wintering areas

Hawk Mountain
Av. fall count: 39

Cape May Point
Av. fall count: 23

St. Louis
Av. fall count: 38

Spring migration sites:

1. Whitefish Point
Av. spring count: 25

2. Derby Hill
Av. spring count: 18

Peak spring migration: March 31–May 31

Holiday Beach
Av. fall count: 21

Hawk Ridge
Av. fall count: 149

Golden Gate
Av. fall count: 1

Goshute Mtns.
Av. fall count: 10

Peak migration: Aug. 20–Nov. 30
A good day's observation: 1

Bob Hines

Sharp-Shinned Hawk
Accipiter striatus

THE SHARP-SHINNED HAWK IS WIDELY DISTRIBUTED ALL ACROSS NORTH America. It is the smallest member of the Accipiters, a group of hawks that feed primarily on other birds. Because of its feeding habits, it often brings out mixed emotions in bird-lovers. On the one hand, it is an exciting bird to see and watch; on the other hand, it may be feeding on birds you are trying to observe or attract.

Of all the hawks in this guide, the sharp-shin is undoubtedly the least well known. There are many reasons for this, including the hawk's small size and its habit of being secretive and staying within the forest canopy. Accipiters in general are also usually shy of humans and urban areas.

Still, the sharp-shin is one of the most commonly seen hawks during fall migration. Populations migrate both along the coasts and inland along mountain ridges. During migration sharp-shins are known for their aggressive behavior. They are so often seen diving upon one another or other hawks that this behavior is deemed a fairly reliable identification clue among hawk-watchers. Why it spends this extra energy while on migration is still a mystery, just as are many of its breeding habits. We hope that, in the coming years, educated behavior-watchers will take a strong interest in figuring out the puzzles of this exciting bird's life.

BEHAVIOR CALENDAR

	TERRITORY	COURTSHIP	NEST-BUILDING	BREEDING	PLUMAGE	SEASONAL MOVEMENT	FLOCK BEHAVIOR
JANUARY							
FEBRUARY							
MARCH		▓				▓	
APRIL		▓	▓	▓		▓	
MAY				▓			
JUNE				▓	▓		
JULY				▓	▓		
AUGUST					▓	▓	
SEPTEMBER					▓	▓	
OCTOBER						▓	
NOVEMBER							
DECEMBER							

DISPLAY GUIDE

Visual Displays

Visual displays have rarely been recorded in the scientific literature. We have seen sharp-shinned hawks do all of the displays mentioned for goshawks, which include: Slow-flapping, Undulating-flight, Tail-flagging, and Dive-display. But until these have also been recorded by other observers we are hesitant to include them in the text. More observations of this aspect of sharp-shin breeding are needed.

Auditory Displays

Kek-Kek-Call
Male or Female   *Sp Su F*
A fairly rapid series of short, harsh, high-pitched sounds. Similar to the Kekeke-call of the Common Flicker. Male version is higher pitched than female version.
CONTEXT: Given by birds during disturbances at nest or near fledged young. *See* Breeding

Peeep-Call
Male or Female   *Sp Su F*
A thin, drawn-out, squealing call.
CONTEXT: Given during food transfer between mates or by young in the fledgling stage. May also be used during courtship. *See* Courtship, Breeding

Two other calls have been infrequently described: a low, soft call given by the male as he arrives near the female with food, and a whinelike call given by the male before copulation.

BEHAVIOR DESCRIPTIONS

Territory

There are no records of territorial defense among sharp-shinned hawks. In general, the birds nest about two to three miles apart and thus have large ranges.

Within their range, the birds will have a nesting area that includes several acres around the nest. Within this area are spots, such as stumps, fallen logs, or rocks, where the birds bring prey items, pluck them of feathers, and eat them. There will also be specific places in the trees where the birds repeatedly roost at night, and later in the season these may be marked by feces and molted feathers. And finally, there are nesting spots, which usually are found in dense evergreens.

The birds return to the same nesting area for several years in succession.

Courtship

Main behavior: Aerial displays, mate-feeding
Duration: From arrival on the breeding ground until mid-nestling phase

There are very few records of sharp-shin courtship. One account describes a pair chasing each other in rapid, erratic flight. Following this, the two birds perched in the same tree and gave the Peeep-call. Gradually, the male approached the female until he was right next to her. He then gave a low whine, stepped onto the back of the female, and they mated for thirty to forty seconds as both flapped their wings. Mating was repeated twice more in the next hour.

The male and female return to their nesting area in spring about two weeks before egg-laying begins. Whether the two arrive together and already paired is not known. Fairly soon after arriving, the male may begin mate-feeding, bringing the female most of her food. As he flies into the nesting area with food he may give a low call. The female may then fly out to him. They land at a perch where he transfers the food to her. She may give the Peeep-call as she takes the food. He continues this mate-feeding through the middle of the nestling phase.

Sharp-shins generally do not start breeding until they are two years old and in their adult plumage. However, there have been cases where both males and females have been seen breeding in their first year, a time when they still have their immature plumage. *See* Molt

Nest-Building

Placement: Usually in dense evergreens next to trunk, 15–60 feet above ground
Size: Outside diameter 20–25 inches, outside depth 5–7 inches
Materials: Thin twigs; lined with fine twigs and flakes of bark

Sharp-shins generally return to the same area each year to breed, but prefer to build new nests. This often results in several nests being located within one to two hundred feet of one another. Sharp-shins prefer to nest in small groves of dense evergreens near clearings. The nests are usually built in trees at the edge of the group. In tall evergreens they tend to nest near the bottom of the canopy, while in smaller evergreens they tend to nest near the top of the trees. In either case, the nest is almost always right next to the trunk.

The nest is started soon after the birds arrive on the breeding ground. It is not known which sex builds most of the nest. Thin twigs are broken off branches for the base of the nest. These are laid in a loose framework, over which is laid a shallow layer of finer twigs and flakes of outer bark pulled off pines, spruces, oaks, and other trees. The nests are usually completed in about a week. Sharp-shins have been known to desert their nest when disturbed during the nest-building phase.

One sign that a nest is in the general vicinity is the presence of collections of feathers on the ground. The birds pluck the feathers from their prey before they eat it. They usually have several plucking perches within the two to three acres around the nest. *See* Territory

Breeding

Eggs: Usually 4 or 5. White to bluish white with darker markings
Incubation: 30–32 days, by female only
Nestling phase: 21–28 days
Fledgling phase: 30–40 days
Broods: 1

Egg-Laying and Incubation

Egg-laying starts soon after the nest is completed. The eggs are laid one every other day until the clutch is complete; thus laying takes seven to nine days. Incubation is believed to start with the laying of the last egg, with the result that the young all hatch within a one-to-two-day period. All incubation is done by the female, and she depends on the male to bring her all of her food.

As a result, the male is rarely near the nest but instead out hunting. As he brings food back he may fly under the forest canopy and give a low call. The female comes off the nest and meets him at a nearby perch to get the food. The Peeep-call may be given by both birds at this time.

Individual birds vary in their response to disturbances around the nest at this stage. The female may fly off silently before you get near the nest and wait quietly nearby until you leave. Other females have been known to stay on the nest until practically touched, but this was observed many years ago when people collected eggs. Collecting eggs is now illegal and we do not recommend disturbing the nest. Other females may fly off the nest and perch or circle overhead giving the Kek-kek-call.

Nestling Phase

The eggs all hatch within about thirty-six hours. Although four or five eggs are generally laid, often only three or four hatch, the other egg or eggs remaining unhatched and eventually being broken by activities in the nest.

The young are covered at first with pure white down. By the first week they have doubled in size. By two weeks their wings are fairly well developed and the young may peer over the edge of the nest. In the last week of the nestling phase they may defecate over the edge of the nest, making whitewash around it. At this point their early down is shed and may cling to the twigs of the nest.

For the first half of the nestling phase the male continues to bring all of the food. It is given to the female, who then takes it to the nest, rips off small bits, and feeds them to the young. During the second half of the nestling phase the female may begin to make some trips to gather food as well. To some extent, how much she participates may be a result of how well the male is doing. If he cannot provide enough food, this may stimulate her to hunt also.

At about three weeks of age, the young stand on the edge of the nest and are just about ready to take their first flights. The nestling phase lasts three to four weeks. In general, the smaller males develop faster and leave after about twenty-four days, while the large females leave at about twenty-seven days.

Fledgling Phase

When the young leave the nest they fly to nearby exposed perches and start to give a call similar to the Peeep-call. They generally stay near one another and near the nest area for about a month. After that the family may drift away from the nest area but still remain together for another week or so.

The young are dependent on the adults for food during this entire period, but near the end the adults feed them substantially less, which forces them to attempt hunting on their own. They may catch insects and try to catch birds, but they are usually not successful for quite a while. They become independent at about the same time passerines (songbirds) start migrating, and this may make it easier for them to survive their first weeks on their own.

The young gradually disperse, leaving one another and their parents.

Plumage

DISTINGUISHING THE SEXES There are no plumage differences between male and female that can be distinguished in the field. However, the female is substantially larger in all respects than the male. In terms of behavior, the female is the one most often near the nest, while the male is the one most often hunting during breeding.

DISTINGUISHING IMMATURES FROM ADULTS Immature sharp-shins have brown backs, heads, tails, and wings, and, underneath, are whitish with heavy brown streaking. They keep this plumage until their first spring, when they begin to molt and acquire the adult plumage. Adult plumage is gray on the back, head, tail, and wings, and whitish underneath with reddish brown barring.

MOLTS Sharp-shins have one complete molt per year. This starts in early summer and continues into early fall.

Seasonal Movement

Sharp-shins breed as far north as the Canadian provinces and Alaska but generally move into the lower forty-eight states and Central America for the winter. They usually migrate in small groups of a few birds or singly, often in the company of other Accipiters or broadwings. Their period of migration often spans a longer time period than that of most other hawks, starting early and continuing through most of the fall.

In general, immatures migrate first, followed by adults. It is also believed that immatures move farther south than adults and that males move farther south than females. This follows a theory that more dominant birds, in this case adult females, migrate the shortest distance.

Spring migration occurs in March and April. It is less concentrated and not as well studied.

RECOGNITION DURING MIGRATION A small hawk with short, rounded wings and a long tail. Most easily recognized by its continual alternation between several flaps of its wings and a single glide: flap,

flap, flap . . . glide . . . flap, flap, flap . . . glide. Because it is a light hawk, it is buffeted about by the wind and rises quickly in thermals, using small circles.

BEHAVIOR DURING MIGRATION Sharp-shins are seen as lone birds or in groups of several birds flying along ridges or coastlines on windy days. They may also be seen inland joining the thermals used by broadwings but, again, usually singly or in twos or threes. Sharp-shins typically attack other sharp-shins and other species of hawks while on migration by diving down upon them from above. Occasionally, sharp-shins catch insects in midair or make a brief chase after a smaller bird close to the ground, possibly trying to catch it for a quick meal.

For more information on hawk-watching, see Appendix B, Hawk-Watching.

Sharp-Shinned Hawk Fall Migration Chart

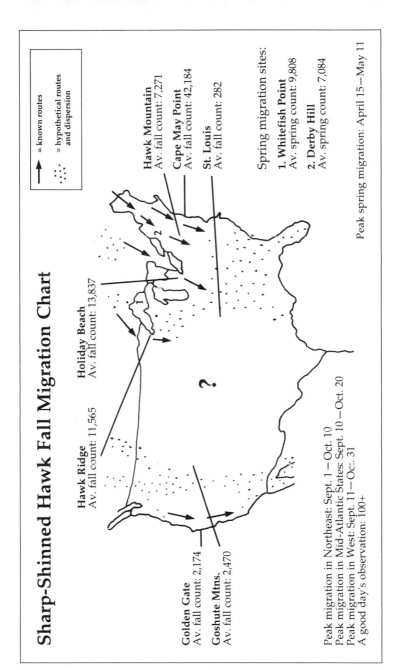

Hawk Ridge
Av. fall count: 11,565

Holiday Beach
Av. fall count: 13,837

Hawk Mountain
Av. fall count: 7,271

Cape May Point
Av. fall count: 42,184

St. Louis
Av. fall count: 282

Golden Gate
Av. fall count: 2,174

Goshute Mtns.
Av. fall count: 2,470

= known routes

= hypothetical routes
and dispersion

Spring migration sites:

1. Whitefish Point
Av. spring count: 9,808

2. Derby Hill
Av. spring count: 7,084

Peak migration in Northeast: Sept. 1 – Oct. 10
Peak migration in Mid-Atlantic States: Sept. 10 – Oct. 20
Peak migration in West: Sept. 11 – Oct. 31
A good day's observation: 100+

Peak spring migration: April 15 – May 11

Bob Hines

Northern Goshawk
Accipiter gentilis

THE NORTHERN GOSHAWK IS A POWERFUL AND THRILLING BIRD TO WATCH. The female is considerably larger than the male and takes the lead in all behavior, seeming to initiate territory formation and court-ship, handling all tasks directly related to the young, and acting as the main defender at the nest. The smaller male spends most of the breeding period hunting for food, which it then brings back for the female to eat or feed to the young.

Thus, with this extreme division of labor, it is rare that you will see the two adults together, even around the nest. The male only brings food three or four times a day and the transfer to the female takes only minutes. Outside of the breeding season all goshawks are solitary.

It is interesting to compare the behavior of the goshawk with that of its smaller cousin the sharp-shinned hawk, which is also covered in this volume. Both have the same division of labor dur-ing the breeding season. They also have similar systems of vocalizations — basically amounting to an alarm and advertising note and a food call. Where they seem to differ most is in visual displays. Goshawks seem to use several aerial displays in the early breeding period, and these are quite often seen by observers. But very few aerial displays have been recorded in the sharp-shin. It makes one wonder whether this marks a real difference in the birds' behaviors or is just a result of the sharp-shin's being smaller, less conspicuous, and less often watched.

BEHAVIOR CALENDAR

	TERRITORY	COURTSHIP	NEST-BUILDING	BREEDING	PLUMAGE	SEASONAL MOVEMENT	FLOCK BEHAVIOR
JANUARY							
FEBRUARY	▓	▓	▓				
MARCH	▓	▓	▓				
APRIL	▓	▓	▓	▓	▓		
MAY	▓	▓		▓	▓		
JUNE	▓			▓	▓		
JULY	▓			▓	▓		
AUGUST	▓			▓	▓		
SEPTEMBER					▓	▓	
OCTOBER						▓	
NOVEMBER						▓	
DECEMBER							

DISPLAY GUIDE

Visual Displays

Goshawks have a number of aerial displays that are done primarily in late winter and spring over the breeding territory. The female takes the lead in these; at times she is joined by the male. The displays may occur in various combinations or appear singly. They may be accompanied by the Keek-keek-call or they may be silent. They are usually done in the morning on clear days. They function to advertise the territory and/or help establish the pair bond between male and female.

Slow-Flapping
Male or Female   *W Sp*
Similar to normal flight, during which glides are interspersed with several flaps of the wings, but in the display the flapping is exaggerated, with the wings going very high and very low with each beat. During the glides the wings are angled up into a dihedral, rather than held flat as in normal flight.
CALL: None
CONTEXT: Generally done over the breeding territory early in the season and in the morning. *See* Courtship

Undulating-Flight
Male or Female   *W Sp*
The bird moves in an undulating flight path, rising up then gliding down repeatedly.
CALL: None or Keek-keek-call

CONTEXT: Done over the breeding territory early in the season and in the morning. Often done close to treetop level. *See* Territory, Courtship

Tail-Flagging
Male or Female   *Sp Su F W*
Bird fluffs out the white feathers under the base of the tail, making a conspicuous white patch.

CALL: None or Keek-keek-call

CONTEXT: May be done while perched or flying. Often done while soaring quite high in the air. Bird may be alone or with its mate. *See* Courtship

Dive-Display
Male or Female   *W Sp*
While soaring over the territory the bird may make a steep dive toward the nesting area and then rise up at the end.

CALL: None or Keek-keek-call

CONTEXT: Done along with other aerial displays. *See* Courtship

Auditory Displays

Keek-Keek-Call
Male or Female   *Sp Su F W*
A short, harsh call repeated in a long series at a rate of four or five per second. Sometimes described as a shrill chatter. This call is weaker and slower when given by the male.

CONTEXT: Used in a variety of contexts. During the breeding season given as a contact call between the pair in the early morning. Used as possible advertisement call during aerial displays. And used as an alarm note upon the approach of intruders or danger. *See* Territory, Courtship, Breeding

Keeaah-Call

Male or Female   *W Sp Su*

A wailing, drawn-out, downward-slurred call, one to two seconds long and repeated at one-second intervals. First rises in pitch and then descends. Sometimes described as a scream. Sometimes given in a shorter version by the female.

CONTEXT: Given mostly by the female just before and just after the male transfers food to her. May be given by male as he approaches with food. Given by developing nestlings and fledglings when they seem hungry. *See* Courtship, Breeding

Guk-Call

Male   *W Sp Su*

A short, clucking sound, similar to the sound made by clicking your tongue against the roof of your mouth.

CONTEXT: May be given by the male as he enters the territory with food for the female. May also be given when he is near the female. *See* Courtship

BEHAVIOR DESCRIPTIONS

Territory

Type: Feeding, nesting, mating
Size: 1–2 square miles
Main behavior: Aerial dives
Duration of defense: During breeding season

Breeding goshawks tend to prefer large areas of deep woods interspersed with clearings, such as fields or swamps. Outside the breeding season the birds wander widely, moving over an area of twenty to forty square miles. Goshawks are very solitary birds at this time and rarely come into contact with one another; even mated pairs live separately in winter. No territories seem to be defended in winter.

Pairs return to previous breeding sites in successive years. They may return up to five months before egg-laying begins, but more generally arrive one to two months in advance. At this time they become much more restricted in their movements, generally staying within an area of about one to two square miles.

The female is believed to be the first to return and may take the lead in advertising and defending the territory that contains the future nest site. She may be seen in the mornings doing any of the aerial displays just over the treetops and giving the Keek-keek-call.

During the breeding season goshawk pairs are usually widely dispersed; therefore territorial interactions are hardly ever seen. If another female should venture near, the territorial female will attempt to soar above it and dive down upon it.

A more common feature of her defense of the territory is seen when humans approach the nest. Female goshawks, in some cases, are known to be relentless in their attacks upon human intruders. They will give the Keek-keek-call and repeatedly make dramatic dives, possibly even hitting the person with their feet or actually scraping them with open talons. Therefore, it is wise to take a territorial female goshawk seriously and keep your distance from the nest site. The birds seem to be less aggressive when there are several people together, and not all females will attack intruders; some just call and circle about.

The male goshawk does not generally take part in territorial defense and may just silently fly off at your approach. It may also occasionally call and watch from a distance.

Within the territory and, generally, within a hundred yards of the nest are several plucking perches. These are arched-over trees or fallen logs or stumps where the male or female plucks the feathers from avian prey before eating it. The plucking perches are used most early in the season. By the time the nestlings hatch, the male may be gathering more songbird nestlings as prey to feed its young and mate, and these do not need plucking.

Courtship

Main behavior: Aerial displays, Keek-keek-call, mate-feeding
Duration: From arrival of male until mid-nestling phase

Northern goshawks can breed in their first year, but most seem to wait until they are two to three years old before starting. After breeding, mates become solitary for the winter and then, in the spring, probably re-establish bonds for the next breeding season.

It is usually the female that arrives on the territory first, and she may begin doing aerial displays over it. These advertise her presence to both neighbors and her mate. She may do any one of the aerial displays listed in the display guide. These generally are done in the morning.

Another common female behavior involves perching on conspicuous sites and giving long bouts of the Keek-keek-call. While perched she may do the Tail-flagging display, in which her white rump feathers are fluffed out.

When the male arrives, he joins the female in aerial displays and the two may engage in chases and dives. At night, male and female roost about fifty to a hundred yards apart on the territory. In the morning, they often exchange Keek-keek-calls in a sort of duet as they start the day.

As the time for egg-laying draws nearer the male begins to do mate-feeding, bringing the female all of her food. During this time she remains in the area of the nest and the male flies off, often venturing far from the nesting area. He gives the Keeaah-call as he re-enters the territory with captured prey. The female usually answers him with the same call and flies over to him. She takes the food from him and flies to a perch to feed. He flies off to do more hunting. Mate-feeding will continue into the second half of the nestling phase.

During the later stages of nest-building, often during the actual building, the female initiates copulation by perching on a branch near the nest, drooping her wings, tail-flagging, and lifting her tail. The male may then droop his wings, tail-flag, and, with a swooping flight, land on her back. They then copulate for about ten seconds, with both birds flapping their wings. Copulation can occur several times a day for the week or two before egg-laying.

In all interactions between the pair during the breeding season, the female seems clearly dominant.

Nest-Building

Placement: Where a tree trunk forks in at least three directions to provide support, 30–60 feet above ground
Size: Large bulky nest; outside diameter 30 inches; outside height 10 inches, or more in later years
Materials: Large branches and sticks; lined with smaller branches, possibly bark chips; green sprigs added through incubation and nestling phase

Goshawks tend to reuse the same nest from year to year. In some cases, a pair may have two nest sites that they have used over the years and may move from one to the other in succeeding years.

Nest-building can be a very leisurely affair for goshawks, since the birds arrive on the breeding territory several months before egg-laying and possibly do some building for up to two months before egg-laying. In general, the female does all of the building, gathering sticks from trees or off the ground and then taking quite a bit of time to place each branch in the nest. Even if the nest is reused, the female will add more material, making the nest become very tall over the years. Sometimes old nests get so large that they are susceptible to being blown down in storms. In these cases, the birds will build a new one in the same spot or a new location.

Nest-building usually takes place in the morning after the birds have done some reciprocal calling. During nest-building, especially in the two weeks before egg-laying, the female initiates copulation nearby. After nest-building, the birds may do some aerial displays over the territory.

Like many hawks, the female frequently collects green sprigs from conifers or other trees and drops them on the nest. She does this most during incubation and the first two thirds of the nestling phase; after that it is done less often. No effort seems to be made to place the sprigs into the nest structure. Rather, they are usually just placed on top. This suggests that they have functions other than providing additional support for the nest. Theories about the purpose of the sprigs range from their providing added moisture to their controlling parasites to their actually serving as food.

Breeding

Eggs: 2–5, average 3 or 4. Whitish to pale blue, usually with no markings
Incubation: 35–38 days, by female only
Nestling phase: About 5 weeks
Fledgling phase: 5–6 weeks
Broods: 1

Egg-Laying and Incubation

The eggs are laid at two-to-three-day intervals and the average clutch contains three or four eggs. During the egg-laying period the female may rest over the eggs to keep them from getting too cold, but full-scale incubation probably does not start until the laying of one of the last two eggs.

The female is dominant around the nest site at this time and is the one who handles all care of the young. The male's main task at this time is to bring food to the female. After transferring food to her away from the nest, the male may briefly go to the nest and stand over the eggs. But the female usually returns in a few minutes, often doing the Tail-flagging display and giving the Keeaah-call, which makes the male leave and resume hunting. Occasionally, the female will leave the eggs to gather green sprigs, which she then drops on the nest. Both birds develop regular patterns of flight that they repeatedly use in the nesting area, always approaching and leaving by the same routes.

The female remains on the eggs at night. The male usually roosts several hundred feet from the nest but in the territory.

Incubation lasts from thirty-five to thirty-eight days.

Nestling Phase

The young hatch over a period of two to three days or less. They are brooded and fed only by the female. The male flies into the territory giving the Keeaah-call, and the female flies out to meet him with the same call. She takes the food from him and again gives the Keeaah-call as she returns to the nest. The male resumes hunting. The female tears off bits of meat and the young peck it out of her beak. If the young no longer beg for food and there is still prey left, the female will carry it away from the nest and cache it in the crotch of a branch. These remains are used later if the young become hungry before the male returns again with more food. Caching is most common in the early weeks of the nestling phase; later, the young can consume larger quantities.

The young are brooded almost continuously for the first week. For the second week they are brooded less during the day but still brooded all night. As of the beginning of the third week they are no longer brooded during the day or night, but the female will come and stand over them during heavy storms. When not brooding, the female perches on the rim of the nest or on nearby branches.

At the beginning of the fourth week, the young may start to give a version of the Keeaah-call when parents arrive with food. At this time they can also feed themselves and whole prey may be dropped off at the nest. In this case, one of the fledglings usually claims a prey item, feeding alone on it and defending it from the others. The young can become quite aggressive toward one another at this time, especially in the absence of the female. Sometimes, when there is a shortage of food, one of the young may be harassed by the others and die. It may even be eaten by the remaining young.

During the last half of the nestling phase, the female spends more time off the nest, perched up to several hundred feet away. At this time she may also start hunting and bringing food to the young. How soon in the nestling phase she starts hunting may be determined by how successful the male is at bringing food. If he cannot gather enough food, this may stimulate her to hunt. When she does hunt it is usually in the general vicinity of the nest, not as far away as the male goes. When the male arrives with food and the female is not immediately present, he may fly to the nest and drop off the food. But as soon as the female shows up, he leaves, possibly prompted by her Keeaah-call.

The young birds in the later weeks of the nestling phase excrete their droppings over the nest edge and these may be seen on the ground below.

Fledgling Phase

At the end of the nestling phase the young may move out of the nest onto nearby branches. Gradually they take flight and move to other trees around the nest. They usually remain within two hundred feet of each other and give loud Keeaah-calls when they are hungry. Both parents will bring food to them at this time. It may be taken to the individual fledgling, dropped off at the nest, or even transferred in midair.

The family stays together over a period of four to five weeks, sometimes until September. They move slightly from day to day, which may carry them quite far from the nest over the weeks. At some point the parents probably begin to feed the young less, which forces them out on their own. By mid-fall the family is dispersed.

Plumage

DISTINGUISHING THE SEXES The plumage of male and female is the same, but the female is substantially larger than the male. She is also the one to take the lead in nest defense and care of the young. The male will be seen hunting or delivering food to the female.

Female calls are, generally, more forceful sounding than those of the male.

DISTINGUISHING IMMATURES FROM ADULTS Adults have gray backs and fine gray barring over their underparts. Immatures' plumage is brown on the back and their undersides are covered with thick brown streaking. This immature plumage is kept until the bird's first spring when it starts to molt and acquire adult plumage.

MOLTS Goshawks have one complete molt per year and this occurs from late spring to early fall. Males generally start this molt later than females.

Seasonal Movement

Goshawks are not generally migratory. Instead, they remain in the vicinity of their breeding ground in winter but wander over a much larger area. In most years, however, some immature birds move south for the winter and these may be seen at hawk migration sites anytime from September to December. These birds move north again in February and March.

There seems to be a cycle of goshawk invasions that repeats approximately every ten years. Thus, once every decade, large numbers of both immatures and adults move south in fall. The cause of this is not known for sure, but the invasion cycles may be linked to population cycles in goshawks that are in turn affected by the population cycles of their prey.

Weather conditions seem to have less effect on goshawk migration than they do on the migration of other hawks. Thus, you may see goshawks migrating under a variety of circumstances and over a long period of time. The birds move singly and not in groups.

RECOGNITION DURING MIGRATION A large, solidly built hawk with compact wings and a long tail. Its proportions and flight patterns are similar to those of its smaller cousin the sharp-shin, but it is clearly a larger and more powerful hawk with relatively longer wings. (It looks like a sharp-shin that has done weight lifting.) It tends to glide more than it flaps and because of this it reminds one of a buteo such as the broadwing or red-tail.

BEHAVIOR DURING MIGRATION Goshawks migrate singly and are often not in the company of other hawks. Because of their powerful flight they may be less dependent on weather conditions for their movement.

For more information on hawk-watching, see Appendix B, Hawk-Watching.

Northern Goshawk Fall Migration Chart

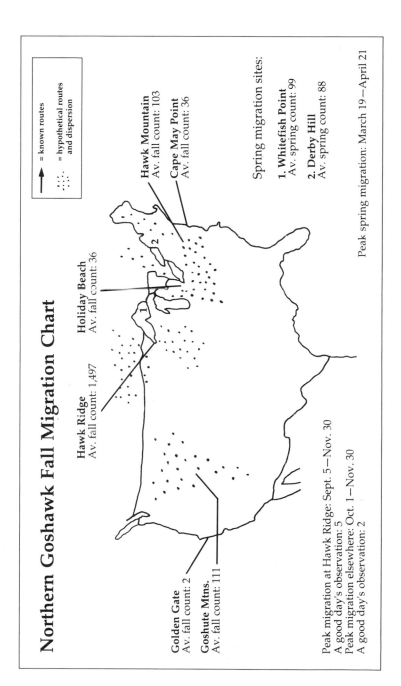

Hawk Ridge
Av. fall count: 1,497

Holiday Beach
Av. fall count: 36

Hawk Mountain
Av. fall count: 103

Cape May Point
Av. fall count: 36

Golden Gate
Av. fall count: 2

Goshute Mtns.
Av. fall count: 111

➤ = known routes

∴ = hypothetical routes and dispersion

Spring migration sites:

1. Whitefish Point
Av. spring count: 99

2. Derby Hill
Av. spring count: 88

Peak migration at Hawk Ridge: Sept. 5—Nov. 30
A good day's observation: 5
Peak migration elsewhere: Oct. 1—Nov. 30
A good day's observation: 2

Peak spring migration: March 19—April 21

Broad-Winged Hawk
Buteo platypterus

ON THE BEST FALL DAYS, WHEN THE AIR IS CLEAR AND A NORTHWEST FRONT is coming in bringing relief from the heat of summer and a hint of the cold to come, broad-winged hawks start their migration. These fall migrations can be spectacular events, with literally thousands of broadwings gliding over certain mountaintops on a single day.

Broadwing migrations have been well studied in the North. What the birds do farther south remains to be discovered by curious hawk-watchers willing to spend time scanning the fall sky and keeping records of the birds' movements. Once the broadwings get to Texas they concentrate in groups of tens of thousands. They can be seen in the greatest numbers in the narrow portions of Central America such as Panama, as they pass through to South America, where they spend the winter.

With all of this interest in their migration, it has been easy to forget that these birds live here in North America all summer as they go through their breeding cycle. Among all our hawks, the broadwing is one of the lesser known in terms of its breeding behavior. On the one hand, the bird is quiet and secretive around its nest, making it hard to discover. On the other hand, it is quite at home near human habitations and there are undoubtedly many nesting pairs that could easily be watched but go unnoticed. The many gaps in our understanding of the breeding behavior of this common bird can only be filled in by knowledgeable observers. The summary that follows will tell you most of what is known about the bird's behavior. To learn more than this you will have to head out on your own and watch the birds for yourself.

BEHAVIOR CALENDAR

	TERRITORY	COURTSHIP	NEST-BUILDING	BREEDING	PLUMAGE	SEASONAL MOVEMENT	FLOCK BEHAVIOR
JANUARY							
FEBRUARY							
MARCH						▓	
APRIL	▓	▓	▓	▓		▓	
MAY	▓	▓	▓	▓	▓		
JUNE	▓			▓			
JULY					▓		
AUGUST					▓		
SEPTEMBER						▓	▓
OCTOBER							
NOVEMBER							
DECEMBER							

DISPLAY GUIDE

Visual Displays

No visual displays of broad-winged hawks have been recorded in the literature so far; however, experienced hawk observers have repeatedly seen at least three actions of broadwings that are probably displays. These are listed below with the caveat that more observations of these actions are needed.

Undulating-Flight

Male or Female   *Sp*

Bird flies high over tree canopy in an undulating flight path and, when flapping wings, uses a stiff motion. Whether this display is done by the male and/or the female is not known.

CALL: None

CONTEXT: Usually seen in spring during the time of migration and the beginning of breeding. May be done in response to other hawks' flying by. May be territorial and/or courtship advertisement. *See* Territory

Pigeon-Flight

Male or Female   *Sp*

Display similar to the pigeon's wing-clapping-flight. Several deep wingbeats are followed by gliding with the wings held over the back in a V. There is no actual hitting of the wings among hawks. Whether this display is done by the male and/or the female is not known.

CALL: None

CONTEXT: May be done between members of a pair just above the tree canopy and near the nest. May be part of courtship. *See* Courtship

Head-Bobbing
Male or Female   *Sp Su*
Bird bobs its head up and down several times in succession. Tail may be spread.
CALL: None or Chitter-call
CONTEXT: A display that may be associated with the transfer of food between mates or even between parents and young. *See* Courtship

Auditory Displays

Sigee-Call
Male or Female   *Sp Su F*
A very high-pitched, thin whistle with a short preceding syllable. The pitch of the whistle is constant or rises very slightly at the end. One second in length.
CONTEXT: The main call of this hawk. Given when you approach the nest, during aggressive encounters with other hawks, and under many other circumstances. Often given when the bird is in flight. Its in-flight function is not understood. Fledglings can give this call for the first time when one month old and begging for food. The female may give a version of this call when she returns to the nest after being fed by the male. *See* Breeding

Whine-Call
Male or Female   *Sp Su*
A series of short, harsh chirping sounds, distinct from the clear whistle of the Sigee-call.

CONTEXT: Given when food is transferred between mates. A more intense version is given
by nestlings when the parents bring them
food. *See* Breeding

BEHAVIOR DESCRIPTIONS

Territory

Type: Nesting, feeding, mating
Size: 1 – 2 square miles
Main behavior: Sigee-call, dives, soaring
Duration of defense: From arrival on the breeding ground until late
fledgling phase

Breeding broad-winged hawks generally return to their previous nesting area in spring and will use the area for the site of a
new nest, breeding, and feeding. They will defend it against other
broad-winged hawks and other large buteos, such as red-tailed
hawks, if the occasion arises, and defense will continue through
the breeding period until the fledglings and parents have left the
nest area.

Broadwings are generally inconspicuous on their territory.
Older pairs may arrive, build a nest, and begin egg-laying within a
week. This does not leave time for elaborate territorial displays.
Also, these displays may not be needed, since the hawks nest far
apart and may be accustomed to their neighbors.

A visual display possibly associated with territory advertisement may be the Undulating-flight that occurs over the territory in
spring just after the birds arrive. It sometimes seems to be done in
response to hawks flying overhead.

Broadwings often nest and hunt in areas where there are also
red-tailed hawks. Sometimes the two get along peacefully, but
there may also be aggressive interactions between the two species.

These involve one or two of each species soaring, giving the Sigee-call, and diving upon one another.

The density of breeding broad-winged hawks depends on the richness of food resources within their nesting habitat. In good habitat — a mixture of woods, fields, and wet areas — the birds can breed within an area of one to two square miles.

In their second year juveniles migrate north a month or more after adults. They do not breed in that year but may wander about. They do not return to the area where they were born, so, if they breed, they must establish territories in new locations.

Courtship

Main behavior: Soaring, Pigeon-flight, mate-feeding
Duration: From arrival on breeding ground until nestling phase

Very little is known about broad-winged hawk courtship in the wild. Soon after the pair arrive on the breeding ground, the male starts to bring food to the female. She may fly out to meet him and either bird may give the Whining-call as food is transferred. Head-bobbing may also occur before the transfer of food.

A pair of broadwings may soar over their territory in what looks like coordinated flight. And the Pigeon-flight may occur in the early courtship of some pairs. Whatever is involved in broadwing courtship, it is brief, since the birds arrive late in the breeding season and start to lay eggs within five to seven days of arrival.

Nest-Building

Placement: Usually in a three- or four-part fork of the main trunk of a deciduous tree, in the lower part of the leaf canopy
Size: Outside diameter 1 – 1½ feet; outside depth 6 inches to 1 foot
Materials: Base constructed of twigs and sticks, lined with bark, moss, and green leaves from evergreen or deciduous trees

Broad-winged hawk nests are usually built in hardwood forests where there are both upland clearings and wooded swamps nearby. In these habitats they can hunt for the mammals, birds, amphibians, and reptiles that they eat. Nests are often near a clearing, foot- or bridle path, or road that may provide a clear path through the woods to the nest. Nesting broad-winged hawks can easily go unnoticed since they are generally quiet and usually approach the nest through the forest rather than above it the way red-tailed hawks do.

The birds tend to return to the area where they nested previously, but do not use the same nest or tree, preferring to build a new nest in a nearby location. However, new nests may be built within fifty yards of previous years' nests.

The hawks often use some existing structure as the base of their nest, such as an old squirrel nest, or old hawk or crow nest. Newly built nests are often larger than nests built on an existing structure. Nests are built quickly and are not particularly well made, so they may blow down during the following winter. The nests are usually placed in the first good crotch of a deciduous tree's upper trunk. They range in height from about ten feet to seventy feet above the ground.

Building begins in April soon after the birds arrive. Both male and female participate in gathering material, although the base structure of twigs and dead sticks may be more the job of the male while the female does more collecting of the lining, which may include bark, moss, live evergreen or deciduous sprigs, and tree flowers. Large twigs are broken directly off trees rather than being collected from the ground. The twigs and sticks are brought to the nest in the birds' feet while lining materials are brought in the birds' bills. The nest is usually completed in several days.

The broad-winged hawk's habit of bringing fresh greenery to the nest, such as tree sprigs with leaves, is common to other hawks as well. Sprig collection starts before egg-laying and continues through incubation and into the last week of the nestling phase. During the nestling phase the female may average three or four sprigs a day. They are taken off trees near the nest. The bird twists them off with its bill and then carries them to the nest in its bill, where they are laid on the rim, in the cup, or, occasionally, on the nestlings. One recent theory on the function of these green twigs is that they may contain chemicals that inhibit the survival of ecto-parasites in the nest.

Breeding

Eggs: Average 2 or 3. Varying shades of white with brownish blotches
Incubation: 30–38 days, by female only
Nestling phase: About 5 weeks
Fledgling phase: 3–4 weeks
Broods: 1

Egg-Laying and Incubation

The broad-winged hawk is one of the latest breeders among the raptors. Egg-laying usually starts at about the time that leaves are beginning to emerge on the trees. One to four eggs may be laid, but more commonly there are two or three in a clutch. They are generally laid one every two days; however, layings may be as much as three to four days apart. The feathers from the female's brood patch start to fall out at this time and may catch on the rim of the nest, giving you a clue to this stage of breeding as you watch from afar. Fresh greenery in the nest is also a good indication that the nest is active.

The female starts to incubate as soon as the first egg is laid. She is the only one to develop a brood patch and the only one to incubate. She stays on the eggs all of the time, getting off only to defecate, gather green sprigs for the nest, or eat food brought to her by the male. During this time the male brings her all her food. As he flies toward the nest she flies out to him giving the Whine-call. Food is transferred at or near the nest, and she takes it to a nearby perch to feed. Meanwhile, the male remains at the nest and possibly stands over the eggs. The female soon returns and gives a version of the Sigeee-call repeatedly until the male leaves the vicinity of the nest. The female then resumes incubation. Before the female settles on the eggs she may reach her head down toward the eggs several times; she may be turning the eggs to aid in their development.

Incubation lasts from thirty to thirty-eight days. If for some reason the eggs are destroyed, the female may lay a replacement clutch within one to two weeks; it usually contains fewer eggs than the first clutch.

The adults are, as noted, generally quiet around the nest, making this phase of breeding inconspicuous. However, individual habits vary. Some birds may give the Sigee-call and fly about when you are near, while others may just fly off the nest and remain hidden.

Nestling Phase

Since the eggs are laid at least a day apart and the female starts incubation with the first egg, the young hatch over a period of several days. The female broods the young continuously for the first week. The male continues his role of providing all food for the female and now the young as well. He does not brood the young and just makes short visits to the nest area to deliver food. The female receives the food away from the nest, takes it to the nest, and tears off bits to feed to the young. The feeding typically takes about ten minutes; if there is still food left, and the young are no longer hungry, the female may go to her perch to eat the remainder. In the first week, the young are only conspicuous when the female is off them and feeding them.

In the second week and later the female generally only broods the young in rain or other inclement weather. She may also do some hunting. The young are more active on the nest, becoming visible as they stand and stretch. In this and the third week the female still has to tear off bits of food for them. By the fourth week the young can eat food on their own and the parents just leave the food off at the nest. The parents bring an average of two prey items to the nest each day.

At this stage the young may begin to fight over food brought. The first hatched is the biggest and usually gets first chance at whatever food there is.

When the young defecate, they back up to the edge of the nest and expel their feces over the edge and onto the ground below. In the later weeks of the nestling phase these feces may accumulate enough to alert you to the presence of the nest. Adults defecate away from the nest.

For the first week the young are downy. In the second week feathers begin to appear, and by the third week the birds are mostly covered with feathers. The nestling phase lasts about five weeks.

Fledgling Phase
By the fifth week the young begin to move out of the nest and onto nearby limbs, but they return to get food at the nest, where the parents leave it off. At this time they can give the adult whistle and do so frequently.

By the sixth week the feathers of the young are fully grown and they can fly. They may get food from the parents away from the nest and may give the Whine-call when receiving it.

After the sixth week neither parents nor young will be seen at the nest. The young will start to hunt on their own in the surrounding area, but will also get some food from the parents. Gradually, the young begin to gather their own food. The family may stay together until very close to the time of migration.

Plumage

DISTINGUISHING THE SEXES There is no way to distinguish male from female through plumage since they look identical; however, the female is slightly larger than the male, weighing about 20 percent more.

DISTINGUISHING IMMATURES FROM ADULTS The most obvious difference is in the markings of the tail and breast. Adults' tails have alternating broad bands of white and black (two of each are usually visible); adults have reddish brown horizontal barring on their breasts. Immature birds (see below) have five or six narrow dark tail bands and brown vertical streaking on their breasts. This plumage is kept until the bird's first spring, when the broadwing starts to molt and acquire adult plumage.

MOLTS Broad-winged hawks have one complete molt per year. This starts in late April or early May and is completed in late August.

Seasonal Movement

Broad-winged hawks migrate each fall to Central and South America and return in the spring. Their fall migration is spectacular since the birds may move in large flocks composed of thousands of broadwings. Typically, the birds rise on thermals, circling about, and then coast in a straight path, falling slightly all the time until they come to another thermal; they then circle again and are lifted on the warm currents.

Broadwings make up by far the majority of hawks seen at

hawk-watch spots, the established areas where people go to see these birds on migration. Thousands of broadwings may migrate over a single mountaintop in one day.

RECOGNITION DURING MIGRATION Broadwings are a medium-sized, compact hawk with broad wings and a short broad tail. Two easy ways to recognize them are: 1) by their habit of soaring up together on thermals with other broadwings, and 2) by noting the broad black-and-white bands on the adult's tail.

BEHAVIOR DURING MIGRATION Almost always seen in groups, rising up on thermals and then peeling off on a glide until they hit the next big thermal. More than almost any of our other hawks, the broadwing depends on the right weather conditions to migrate. In general, good thermals accompanied by a tail- or crosswind are ideal weather for broadwing migration.

For more information on hawk-watching, see Appendix B, Hawk-Watching.

Broad-Winged Hawk Fall Migration Chart

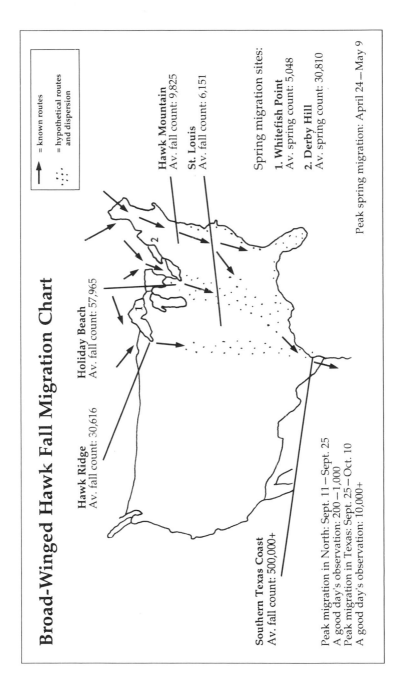

Hawk Ridge
Av. fall count: 30,616

Holiday Beach
Av. fall count: 57,965

Hawk Mountain
Av. fall count: 9,825

St. Louis
Av. fall count: 6,151

Spring migration sites:

1. Whitefish Point
Av. spring count: 5,048

2. Derby Hill
Av. spring count: 30,810

Peak spring migration: April 24 — May 9

= known routes

= hypothetical routes
and dispersion

Southern Texas Coast
Av. fall count: 500,000+

Peak migration in North: Sept. 11 — Sept. 25
A good day's observation: 200 — 1,000
Peak migration in Texas: Sept. 25 — Oct. 10
A good day's observation: 10,000+

Red-Tailed Hawk
Buteo jamaicensis

THE RED-TAILED HAWK IS ONE OF OUR LARGEST BUTEOS AND ON ANY STILL, sunny day, it is likely to be seen soaring overhead. It occurs throughout North America and lives in a wide variety of habitats.

Although it is a common bird, there is still a great deal to be learned about its behavior, especially during the breeding season. It prefers to nest in rural or wild areas and at this time is easily disturbed, often abandoning its nest during nest-building or incubation phases if humans approach too closely. Therefore, some care must be taken in observing red-tails. Stay as far away as possible while still being able to watch through a scope or powerful binoculars. The more you know about the hawk's behavior, the better you can avoid disturbing the hawk's life.

In contrast to this, there are occasionally pairs that nest right in suburban areas, often along roads where there is mostly car traffic and very few pedestrians. As with many birds, you may be able to park your car near a nest, using it as a blind. The birds do not seem as disturbed by the car.

Be sure to look for red-tail aerial activities in late winter and early spring. These are exciting events, part of both courtship and territorial behavior. Watch on a clear day, about mid-morning, from an area where you can see over a large patch of countryside. First one red-tail and then another may rise on the thermals of warm air. If you watch them over a period of fifteen to thirty minutes you are very likely to see the displays of Talon-drop, Undulating-flight, or the Dive-display. They can provide you with a full morning of enjoyment.

BEHAVIOR CALENDAR

	TERRITORY	COURTSHIP	NEST-BUILDING	BREEDING	PLUMAGE	SEASONAL MOVEMENT	FLOCK BEHAVIOR
JANUARY							
FEBRUARY	▓	▓					
MARCH	▓	▓	▓	▓		▓	
APRIL	▓		▓	▓		▓	
MAY	▓		▓				
JUNE	▓			▓	▓		
JULY					▓		
AUGUST					▓	▓	
SEPTEMBER	▓					▓	
OCTOBER						▓	
NOVEMBER						▓	
DECEMBER							

DISPLAY GUIDE

Visual Displays

Talon-Drop
Male or Female   *Sp Su F W*
Bird extends legs down while soaring in the air. It may or may not attempt to touch or hit another bird with its talons.

CALL: Scream or Chwirk-call

CONTEXT: Done during territorial or courtship encounters. More usually done by the male when soaring over the female, but can also be done by female. *See* Territory, Courtship

Undulating-Flight
Male or Female   *Sp Su F W*
Bird in flight starts on an undulating flight path, diving down with wings tucked and then opening wings on ascent, and continuing this for several undulations. Sometimes ends with bird making the Dive-display into woods.

CALL: Scream or Chwirk-call

CONTEXT: This is an aerial display done over the territory. May serve in territorial advertisement and as part of courtship. *See* Territory, Courtship

Dive-Display
Male or Female   *Sp Su*
A steep dive starting from soaring or Undulating-flight and usually heading toward the area of the nest. Stops short of the ground or tree canopy and is not directed at prey.

CALL: None

CONTEXT: Done over the territory and often directed at the nest area. May be part of courtship or territory advertisement.

Auditory Displays

Scream

Male or Female   *Sp Su F W*

A high-pitched, downward-slurred scream, lasting one to two seconds. Sounds like "Tseeeeeeaarr."

CONTEXT: Usually given while in flight, under a variety of circumstances. Mainly directed toward intruders, such as other hawks. Also given when bird is bothered by mobbing crows or by a human approaching the nest. Given while bird is hunting; can also be given during courtship.

Chwirk-Call

Male or Female   *Sp Su F W*

A loud piercing chirp. May be given in a series at intervals of about one second.

CONTEXT: Given while pair are soaring together, either during courtship or after territorial encounters with other hawks.

Kloo-Eek-Call

Male or Female   *Sp Su F W*

A harsh, two-part, upward-slurred call, given repeatedly.

CONTEXT: Given by adults primarily during the breeding season and near the nest. This is also a call given by fledglings in late summer, from the time they leave the nest until they are independent. When the fledgling sees the adult arriving with food, it gives this call with wings flapping loosely at its side. *See* Breeding

BEHAVIOR DESCRIPTIONS

Territory

Type: Mating, nesting, feeding
Size: Average 1.5 square miles but may be larger
Main behavior: Soaring, chasing, giving scream
Duration of defense: All year if owners are not migratory

Red-tailed hawks are generally territorial throughout the time they are on the breeding ground. For resident birds, this may be all year; for migratory birds, this is only during the breeding season.

Red-tail territories range in size from half a square mile to over two square miles, the size determined by the abundance of food and the number of good perches and nest sites. In general, neighboring birds avoid each other; however, when too close, neighboring pairs will interact aggressively, as will a territorial pair and a lone bird, or two lone birds, one of which is claiming a territory. These interactions involve several types of behaviors as well as displays, and although they can occur at any time of year, they are most common in late winter and spring.

One territorial activity is soaring over the territory. A study on the function of soaring in red-tails found that although soaring may be used for hunting, it also plays an important part in territory defense. Through soaring, hawks are able to survey their territory more easily and locate intruding hawks. Soaring occurs most on clear days when the sun warms the earth and creates rising thermals, which the hawks then use for lift. Thermals do not usually occur until about midmorning, so that is one of the best times to watch for red-tail behavior.

Many territorial interactions occur while the birds are soaring.

A typical encounter between two hawks consists of the birds soaring up together, each one seeming to try to get higher than the other. Once they are quite high in the air, the upper bird may do the Talon-drop and then dive down onto the other bird. If the upper bird comes very close, the lower bird may flip over at the last moment and also present its talons. Usually both birds then separate and soar up again, possibly repeating the Talon-drop and dive. The Scream may be given during the interaction. This is similar to some aspects of courtship, except that at the end of territorial encounters one bird leaves the area and soars off far away.

Another behavior believed to be territorial is Undulating-flight. In this exciting display a bird repeatedly dives and rises in a long undulating flight path, occasionally giving the Scream or Chwirk-call at the same time. This is believed to be a territorial advertisement. It may end with the bird doing the Dive-display, in which it tucks in its wings and makes a spectacular dive, often down toward the immediate area of the nest site.

A territorial bird also will attack a perched intruder. Doing a direct dive, and Screaming, it will swoop down and try to knock the intruder off the perch. The attacked bird will flee and may be pursued far beyond the border. The resident will try to rise higher than the intruder and then dive down upon it.

Red-tails have also been known to dive at and chase other species of hawks, such as broad-winged hawks, and even the larger golden eagles and bald eagles. Rarely do they swoop at turkey vultures.

There may also be some territorial activity in the fall. This can be performed by hawks setting up or defending winter territories. They may be defending these from migrating hawks passing through who are also looking for areas in which to spend the winter.

Courtship

Main behavior: Pairs soaring close to one another, Undulating-flight, Talon-drop, Chwirk-call, Scream
Duration: Occurs mainly in spring, but some elements may be seen at any time of year

Pairs of red-tails can remain together for years on the same territory. If one of the pair dies, another hawk will acquire the mate and territory. Courtship activity between red-tails may be seen at any time of year but it occurs most frequently in the spring right before the nesting season.

The pair are not as strongly attached during fall and winter as they are during breeding. One of the first signs of the coming breeding season is the male and female's perching nearer each other, sometimes in the same tree, as they hunt. In general, this close an association between two red-tails indicates that they are a pair. They otherwise would not tolerate each other except under circumstances of extreme crowding due to food shortage.

Among red-tails it is not always easy to distinguish courtship from territorial behavior. During courtship, the two birds soar near each other in wide circles. The circles may get smaller and the birds closer, and sometimes it looks as if they will almost touch or collide. Typically, one bird, usually the male, will soar above and slightly behind the other and do the Talon-drop. Sometimes both members of a pair will do this display at once. They may give Screams or Chwirk-calls during this time, but if the birds are too distant you may not hear this.

These flights can last ten minutes or longer. One bird may also drop slowly down upon the other in a parachute fashion, with the wings lifted above the head and the talons down, until they almost touch. This is similar to an aggressive territorial interaction but the birds do not separate afterward. Instead, both stay on the territory.

During this displaying one of the birds may do Undulating-flight or the Dive-display.

Mating usually takes place after the hawks do aerial courtship maneuvers. The birds will land on a perch and they may mutually preen each other's head and neck feathers. The female postures by tilting forward with wings loosely dangling at her sides. The male will step on her back, perhaps flapping his wings slowly to keep his balance, and copulation takes place, usually lasting five to ten seconds.

Most paired red-tails are adults but there are records of first-year birds breeding with adults.

Nest-Building

Placement: 15 – 90 feet above ground in a deciduous or conifer tree, built near top in crotch of 2 or more large limbs or where limbs meet trunk

Size: Outside diameter 28 – 38 inches. Old nests that have been used for years may be 3 or more feet tall

Materials: Sticks and twigs half an inch or less in thickness, lined with bark of cedar, grapevine, moss, and a few green sprigs of pine, cedar, or hemlock

Red-tails are quite secretive about their nest-building, and often desert their nest if disturbed during this phase. Therefore, it is best to keep as great a distance as is possible from these birds during this time. Try to locate a good vantage point at a distance and watch them through a scope.

The birds may reuse an old nest or build a new one, sometimes using as a base a nest of a squirrel or another hawk. Both birds build the nest. They break off twigs from trees and carry them to the nest in their talons. The freshly broken off ends of the twigs are visible in the nest and a good indication that the nest was recently built. The nest is then lined with shredded bark, pine needles, corn husks, or other material locally available. The birds can complete a nest in a week or less.

From the beginning of nest-building up until when the young fledge, red-tails may deposit green sprigs of leaves or pine needles in the nest. In fact, even before active nest-building begins, the pair may deposit sprigs in several old nests. It is not known for sure why they do this, but there are several theories, one being that it provides the young with clean resting places, since the nest gets quite soiled with pellets and bits of food remains that the young haven't eaten. See chapters on other hawks and eagles for other possible reasons for this habit.

Red-tailed hawks and great horned owls often nest and hunt in similar habitats. The owls do not build nests of their own, and therefore often use red-tail nests. Since they start breeding earlier in the season than red-tails, they may claim the red-tail's previous year's nest. When the red-tail starts to breed, there is no competition over nests with the owls — when a previous nest is occupied by an owl, the red-tail simply builds a new one. In some cases, great horned owls and red-tails may alternate use of the same nest in successive years.

In autumn, the birds may also do some repair to a nest on their territory.

Breeding

Eggs: 1 – 5, usually 2 in the east, 3 – 5 in central or western areas. Bluish white with brown splotches
Incubation: 28 – 35 days, by both sexes
Nestling phase: 44 – 46 days
Fledgling phase: 4 – 6 weeks
Broods: 1

Egg-Laying and Incubation

There may be a substantial period of time between the completion of the nest and the laying of the first egg — possibly as much as four weeks. This time is much shorter in the North where, since the birds arrive on the breeding ground later and leave earlier, they have less time to complete breeding. Very little is known regarding the intervals between the female's egg-layings; however, from the behavior of other large raptors, we might assume that the interval between layings is two or more days. It is also not known whether incubation starts right after the first egg is laid. It probably does, since the birds start breeding early in the season when the temperatures are often still quite low.

Both male and female incubate and incubation lasts for four to five weeks. The male may bring food to the female while she is incubating, but this mate-feeding is not as frequent as in other hawk species, such as ospreys. In red-tails, it seems that the roles of the sexes in raising the young are more alike.

During egg-laying and incubation, red-tails may be quite sensitive to disturbances, such as your own approach to the nesting area. In some cases, the birds seem quick to abandon the nest when disturbed at this time. Therefore, we recommend that you keep as far away as you can from a pair in the early breeding stages. Again, watch only from a distance, possibly using a scope. A typical reaction of the birds to your approach at this time is to fly off the nest, give the Scream, and then circle overhead.

Nestling Phase

The young hatch at intervals of one to two days. When first hatched, they are covered with white downy fluff. For the first several days they may move about the nest, peeping and bouncing up and down. The adults bring food to the nest and tear off bits to feed to the young. At this stage, uneaten food is carried away from the nest by the adults. Later in the nestling phase it may be left in the nest, resulting sometimes in a considerable accumulation. Sometimes one of the young may die for one reason or another, and the body is generally removed by the adult.

At six to seven days, the young will peck at food brought to the nest by the adults and attempt to feed themselves. After ten days, the young will give a high whistling note when an adult soars overhead. At sixteen days they begin to grow feathers.

When the young are four to five weeks old, the adults may just leave off food at the nest. The young peck off the flesh of the prey from the bones, which at this point are not removed from the nest. The young are more active on the nest, walking about the rim, flapping their wings, and calling as the parents approach. The young back up to the edge of the nest to void their excrement, and this creates a ring of whitewash on the branches and ground below. This is a good clue to the presence of an active nest and indicates the later stages of the nestling phase have begun.

During the later weeks of the nestling phase the young are apt to walk out on limbs adjacent to the nest. At about six to seven weeks after hatching the young can fly and leave the nest.

Green sprigs of pine or other trees continue to be brought to the nest by the adults throughout this phase.

Some major causes of nest failure include predation by raccoons and violent storms that may harm the nest, the eggs, or the young.

There have been several recorded instances of three adult red-tails attending a nest. In one case, one male and two females attended a nest containing one young. Both females fed and brooded the nestling but the male was not observed feeding the nestling. In this case it was not known if the male had mated with both females, or if the second female was a nonreproductive helper at the nest.

Fledgling Phase

The first flights of the young can be quite strong and may carry them over a hundred yards from the nest. Once out of the nest, the young begin to give the Kloo-eek-call, which is loud and piercing, making this the most conspicuous phase of red-tail breeding. If red-tails have nested anywhere near your home, this constant calling of the young can become quite tiresome. It intensifies when the adults are seen arriving with food.

For the first two weeks, the young remain quite stationary; they may stay in the same tree for several days. They may occasionally use the nest as a feeding platform. After about three weeks they fly about more and drift away from the nest area. After four weeks they have to do more feeding on their own, since the parents taper off their food deliveries to the young.

At this stage the young continue, though much less frequently, to give the Kloo-eek-call, even though they are feeding on their own. These immature birds may remain in or close to their parents' territory for the remainder of the summer and into fall. If they wander into another pair's territory they may be attacked. Thus, these immature birds become quite secretive and quiet. They often stay at the edges of territories where they may go unnoticed and they keep their soaring to a minimum, possibly because it would make them conspicuous. At some point they must search for an area of their own.

During winter these immature hawks may be displaced by adults from good feeding areas and forced into more marginal habitats. They may be forced into more urban areas than the adults, or forced further south. Some immatures may try to defend feeding territories from other red-tails in fall and winter.

Plumage

DISTINGUISHING THE SEXES There are no reliable plumage clues to the difference between male and female, and of all of our hawks, the red-tail is one of the least dimorphic in size. Even through behavior there are few ways to distinguish the sexes. In general, the female is slightly larger, spends more time on the nest during the incubation and nestling phases, and is the more aggressive bird around the nest. The male, on the other hand, may be more likely to initiate territorial defense in the larger sense.

DISTINGUISHING IMMATURES FROM ADULTS Immature red-tailed hawks are best distinguished from adults by their tails. The immature has a gray-brown tail with many horizontal dark bands.

POLYMORPHISM The red-tail varies in color, depending on the area of the country in which it is encountered. There is a light and a dark or melanistic phase, with some birds intergrading between them, as well as several subspecies.

MOLTS Red-tailed hawks have one complete molt per year, starting in early summer and continuing into fall.

Seasonal Movement

There are basically two types of seasonal movement among red-tailed hawks. One is spring and fall migration; the other is the dispersal of immature hawks immediately after the breeding season.

Immature red-tails become independent from their parents in mid- to late summer and may drift out in all directions, even northward, from the area of their breeding. Which way they move may be determined in part by prevailing winds that carry them in a general direction. This movement of immature red-tails continues through late summer and early fall.

Red-tail migration starts in mid-fall. Not all red-tails are migratory. Those in the far North are all migratory; those in middle latitudes may move slightly south, depending on weather conditions and the availability of food; those in southern states are usually year-round residents.

The first northern hawks to move south are the immatures, possibly including many that dispersed north earlier in the summer. These are then followed by the adults. Immature migrants tend to move further south than adult migrants.

Spring migration occurs in early spring.

RECOGNITION DURING MIGRATION Red-tailed hawks are large, soaring birds. Their wings are typically held out in a shallow V or dihedral. As the bird circles you may be able to see the reddish brown tail of the adult. Both young and adult have a distinctive black leading edge to the inner part of their wings.

BEHAVIOR DURING MIGRATION Red-tails move singly during migration, often in the company of broadwings but rising more slowly and in larger circles on the thermals due to their larger size and greater weight.

For more information on hawk-watching, see Appendix B, Hawk-Watching.

Red-Tailed Hawk Fall Migration Chart

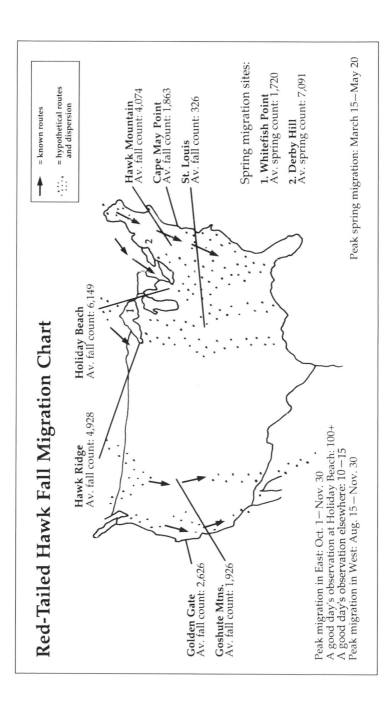

→ = known routes

⋮⋮⋮ = hypothetical routes and dispersion

Hawk Ridge
Av. fall count: 4,928

Holiday Beach
Av. fall count: 6,149

Hawk Mountain
Av. fall count: 4,074

Cape May Point
Av. fall count: 1,863

St. Louis
Av. fall count: 326

Golden Gate
Av. fall count: 2,626

Goshute Mtns.
Av. fall count: 1,926

Spring migration sites:

1. Whitefish Point
Av. spring count: 1,720

2. Derby Hill
Av. spring count: 7,091

Peak migration in East: Oct. 1—Nov. 30
A good day's observation at Holiday Beach: 100+
A good day's observation elsewhere: 10—15
Peak migration in West: Aug. 15—Nov. 30

Peak spring migration: March 15—May 20

Osprey
Pandion haliaetus

AMONG OUR BIRDS OF PREY THE OSPREY IS ONE OF THE MOST AMENABLE TO living near humans. Its main requirements are open water where it can hunt for fish and a platform or strong tree where it can build its nest. Ospreys have occasionally built nests right next to homes, in parking lots, and in public parks. Although they do not prefer being near humans, they do seem to tolerate human presence, an ability that is a big asset for the survival of any species.

Even so, in the 1950s and 1960s the osprey population began to decline in the Northeast. Nesting success was very low and colonies that had once contained over a hundred nests were largely empty. It was believed at the time that these failures resulted from DDT derivatives that were ingested by the fish the hawks ate. Subsequent controls on pesticides may have aided the recovery of osprey populations in the Northeast. Human-made nesting platforms for ospreys in favorable habitats also helped the birds raise successful broods; this was important since many of the trees that the birds would use naturally have been cut down to make way for development.

The osprey is a wonderful bird for behavior-watching. Its calls are distinctive and can help alert you to interesting behavior at the nest. As with all birds, it is important not to approach the nest so closely that you disturb the birds and make them fly up. Just keep your distance and enjoy the constant food trips by the male, the guarding of the nest by the female, the continual bits of nest-building, and the gradual development of the young.

BEHAVIOR CALENDAR

	TERRITORY	COURTSHIP	NEST-BUILDING	BREEDING	PLUMAGE	SEASONAL MOVEMENT	FLOCK BEHAVIOR
JANUARY							
FEBRUARY							
MARCH	▓	▓				▓	
APRIL	▓	▓	▓	▓		▓	
MAY	▓	▓	▓	▓			
JUNE	▓			▓			
JULY	▓			▓			
AUGUST	▓			▓		▓	
SEPTEMBER						▓	
OCTOBER						▓	
NOVEMBER							
DECEMBER							

DISPLAY GUIDE

Visual Displays

Sky-Dance
Male   *Sp Su*

Bird flies sharply up with rapid wingbeats, possibly carrying a fish or nesting material. After reaching a height of several hundred feet, it hovers with tail fanned and talons dangling down. Then it dives down for a varying distance and swoops up again to repeat the hovering. The display may be repeated several times and last several minutes.

CALL: Creee-call

CONTEXT: Usually given over the territory or specifically the nest site. Starts upon first arrival of the male and is most often given on clear days. It is done less frequently after incubation starts. Occasionally given over fishing area late in season. *See* Territory

Hover-Flight
Male or Female   *Sp Su*

Bird hovers while dangling talons; in the case of the male, the bird may be holding a fish. Does not include the steep ascent or dives of the Sky-dance.

CALL: Chirp-call or Chirchirchir-call

CONTEXT: Given by male or female when there is a disturbance at the nest. *See* Territory

Auditory Displays

Chirp-Call
Male or Female *Sp Su*

A single, down-slurred, short chirp. Given in a series of four to ten calls with middle repetitions often slightly higher-pitched than others. Given slowly, at rate of about two per second.

CONTEXT: Given mostly by female around nest when intruding ospreys or other large birds fly near. May be used by males when hunting.

Chirchirchir-Call
Male or Female *Sp Su*

A high-pitched rapidly repeated note given in a series of four to eight repetitions per call. Higher and faster than the Chirp-call.

CONTEXT: Given by male or female when mildly disturbed around nest; may get higher-pitched and faster with increased excitement. Similar call given by female as male arrives with food at the nest.

Grating-Call
Male or Female *Sp Su*

A grating percussive sound. Very un-birdlike and strange. Similar to Chirchirchir-call in rhythm but without pitch—sounds something like "grrgrrgrrgrr."

CONTEXT: Seems to be a gradient of the Chirchirchir-call, possibly expressing greater alarm.

Creee-Call
Male or Female   *Sp Su*
A single, drawn-out, ascending squeal.
Sometimes repeated many times.
CONTEXT: Given in intense alarm situations
and during Sky-dance. Directed at ground or
aerial predators.

BEHAVIOR DESCRIPTIONS

Territory

Type: Nesting, mating
Size: Area right around the nest, occasionally more
Main behavior: Sky-dance, Hover-flight, calls
Duration of defense: Throughout breeding season

Ospreys return to the general vicinity of their birth in their second year, their first year being spent entirely on their wintering grounds. At this time they may pair with another bird and even start to build a nest, but no further breeding activity occurs in that year. In their third year and from then on, they return to this nest site to breed.

Sky-dance early in the season when the male first arrives may be a form of territorial advertisement, for it usually takes place right over the nest site.

Nests can be widely dispersed or as close as fifty feet apart, depending on the availability of trees and crags to support the nests. Both birds defend the immediate area around the nest, but even more so the female, since the male is often out fishing for her and the young. Female defense involves her spreading her wings out over the nest and calling, circling over the nest while giving calls, and doing the Hover-flight with or without nesting material.

Male reaction to an intruder at the nest site includes flying after it as it leaves the area or doing the Sky-dance with or without fish or nesting material. Nest defense continues from first arrival on the territory into August and the end of the breeding season.

In areas where there are many pairs of ospreys, birds are frequently flying over others' nest sites and this often causes mild alarm in the nest owner. Greater alarm is caused when other ospreys try to intrude on the nest, to mate with the female, or to appropriate the nest site for themselves.

Ospreys also respond to other species that intrude around the nest, especially bald eagles, which may be followed by the osprey until they are well away from the nest, with the osprey diving onto the back of the eagle repeatedly.

Courtship

Main behavior: Sky-dance, mate-feeding
Duration: From arrival on breeding ground through incubation phase

Male and female presumably winter separately and meet back on the breeding ground due to territorial fidelity. They arrive at the nest site within a few days of each other, usually the male first. The pair may circle about together and even occasionally dive after one another in the air.

The major part of courtship behavior may be the repeated Sky-dance of the male over the nest site. The male osprey starts these displays as soon as he arrives and they become more frequent over time until the female arrives, then less frequent up until the time of incubation. These may have a double function of advertising for a mate and announcing territorial ownership.

Before incubation, paired birds spend about a quarter of their time at the nest site together. During this time you will see two other aspects of osprey courtship: mate-feeding and copulation. Copulation occurs very often prior to incubation, as many as fifteen to twenty times per day for a three-week period. It is often started after the male returns with food or nesting material. The female may do a display with wings slightly drooped, body horizontal, and tail up, or the male may turn his back to the female, spread and droop his wings, and spread and depress his tail. Copulation may then follow with the female bending forward and lifting her tail to one side and the male alighting on her back with talons closed into a fist. They mate for ten to twenty seconds and then the male gets off. There may be variations in this behavior from pair to pair, with the male possibly displaying to the female as he walks about her.

Mate-feeding is also common, since, soon after the male and female arrive on the breeding ground, the male starts to bring all the female's food to her while she remains in the area of the nest. The male generally returns to the nest with all of his food as well. When the female sees him coming she begins giving the Chirchirchir-call.

The male may stop on a perch within the territory and feed some on the head and front portions of the fish and then fly to the nest, where the female takes the remaining portion. She generally flies to a nearby perch to eat. The male brings about two or three fish per day to the nest prior to incubation.

Nest-Building

Placement: In the tops of trees, large forking branches, or ledges; near water
Size: 5 feet in diameter, 2–7 feet in height
Materials: Large branches; lined with moss, bark, grass, twigs

The nest is built in a spot with good visibility in all directions, often at the crown of a tree or on strong lower branches. It may also be built on rocks, buildings, buoys, electric towers, or human-made platforms. Rarely is it built on the ground.

Nests may be built singly, or there may be several close together in a loose colony. Often the presence of one nest seems to attract other ospreys to nest nearby. The number of nests in a locality is generally related to the abundance of food in the area.

Two-year-old birds in the breeding area for their first time may build a nest but do not lay eggs. Birds who nest unsuccessfully at one spot may build another nest elsewhere later in the season and then return to that spot the next year. Older, successful pairs return to nests used in previous seasons.

Birds start to build soon after they arrive on the breeding ground. The male does most of the collecting of material for the main structure of the nest while the female adds material to the lining. A completely new nest can be built in seven to ten days. Usually an old nest is renovated — to what extent depends on its condition. Branches for the platform are about 20 inches long and are collected not from the ground but from dead limbs on trees. The bird lands on the branch near its end and its weight breaks the twig; the bird then carries the branch back in its feet. Branches may also be broken off as the bird grabs them in flight. Sticks and other nesting material that has fallen down beneath the nest may be picked up and reused. The female brings moss, bark, grass, or small twigs for the lining.

Building may start up to three weeks before egg-laying. From the start of incubation on through the nestling phase the female continues to bring material to the nest.

The average nest diameter is about 5 feet, with a depth of 2 to 3 feet, but if used for many years the nest may be up to 6 to 7 feet deep; the diameter of the inner cup is about 2½ feet.

Breeding

Eggs: 2 or 3. Whitish with reddish brown blotches
Incubation: 34 – 40 days, by male and female
Nestling phase: 7 – 8 weeks
Fledgling phase: 4 – 8 weeks
Broods: 1

Egg-Laying and Incubation

Birds generally lay their first clutch when three years old and this usually contains only two eggs. Clutches in the following years usually contain three eggs. The eggs are laid one to three days apart.

Incubation begins with the laying of the first egg and at this time most mating ceases. The incubating bird sits very low on the nest. At night, the female does all of the incubation. During the day she also does most of the incubation, except when the male comes to the nest with food for her. At this time she gets off the nest and takes the food to a nearby perch to feed while the male takes over incubating the eggs. When finished eating, she returns to the nest and the male leaves, either perching nearby or going off to hunt. Throughout this period the male gets all the food for the female and himself. While incubating, the female may leave the nest briefly to defecate.

Sometimes the male comes to the nest with nesting material and the female then leaves. She may return with lining material as the two trade places again.

Before egg-laying, the two birds roost at night on perches near the nest; after egg-laying the female remains on the nest incubating at night, while the male roosts nearby. Incubation lasts thirty-four to forty days.

Nestling Phase

The young hatch over several days and this results in the first hatched being larger than the last hatched. During breeding seasons when there is a lack of food, there may be aggression between the young, with the stronger and older young getting the most food and having the greatest chance to survive.

For the first ten days the chicks are brooded constantly by the female. The male does all of the hunting and brings the fish to the female, who then tears off bits and feeds it to the young, continuing to feed them this way for the first six weeks.

The female remains on the nest for the first three to four weeks, occasionally covering the young with her wings to protect them from intruders and extreme cold or warmth. After three to four weeks the young begin to exercise their wings, flapping them while holding on to the edge of the nest. At this point, the female may move to a nearby perch to guard the nest.

After the sixth week the female may leave the nest for brief periods to hunt and bring back additional food. These are her first food trips since her arrival on the breeding ground. Fish are now dropped off at the nest and the young feed themselves, with usually only one young eating a given fish.

The nestling phase lasts seven to eight weeks.

Fledgling Phase

After taking their first flight the young continue for a month to return to the nest each night to roost. During the day they usually perch near the nest or near the male's feeding perch, paying constant attention to him and calling when he returns with food. Two weeks after fledging, they may begin to follow the male on his hunting trips and perch nearby giving constant food calls. Not until one or two months after fledging do the young catch fish on their own. They seem to be instinctively able to do this and need no "training" from the parents.

Plumage

DISTINGUISHING THE SEXES This is difficult when seeing an individual bird but easier when looking at a breeding pair. The female is slightly larger than the male and may have a more pronounced dark band across her upper breast. In terms of behavior, the male brings all of the fish to the female during the preincubation, incubation, and early nestling phases.

DISTINGUISHING IMMATURES FROM ADULTS Immatures are similar to adults in plumage but lighter brown on their back; also, all their dark feathers are conspicuously edged with a light beige unlike the all-dark feathers of the adult. Tail feathers are more conspicuously barred. Young birds are indistinguishable from adults when about eighteen months old.

MOLTS The initial molt starts in January or February of the first year and is not totally completed until the bird is about five years old. The next molt begins before the first is completed. Molt cycles take less time to be completed as the bird gets older. Most molting takes place in summer and winter and is arrested during spring and fall migrations.

Seasonal Movement

Ospreys migrate singly and along a broad front rather than in groups and along specific routes as is the case with many other hawks. Spring migration starts in March with birds arriving on the breeding ground in late March and early April. One-year-old (immature) birds remain on the wintering grounds for their first summer and do not migrate north. Two-year-old (adult) birds migrate north, perhaps later than older adults, returning to near where they were born. Three-year olds and older birds return to previous nesting sites.

Fall migration for immatures may start in late August. Migration of adults starts in September and some migration continues into October. Seventy-five percent of all ospreys have migrated by the end of September.

Ospreys from the East Coast migrate along the coast and inland, crossing the ocean along the West Indies and wintering in northern South America. Ospreys from central and western North America migrate south through Central America and into northern South America. A few individuals may winter in the West Indies and southern Central America but most seem to winter along the coast and on the inland waterways of Colombia and Brazil. On their wintering ground they occasionally have been seen roosting in small groups.

RECOGNITION DURING MIGRATION Ospreys are large hawks with long, thin wings that have a characteristic bend in them that gives them a shallow M shape similar to that of a gull's wings. They can be distinguished from gulls by their relatively small head and long tail.

BEHAVIOR DURING MIGRATION Ospreys migrate low or high depending on the weather conditions and are generally alone or using thermals along with other species of hawks. Occasionally, they may fish while on migration as they pass a river or lake.

For more information on hawk-watching, see Appendix B, Hawk-Watching.

Osprey Fall Migration Chart

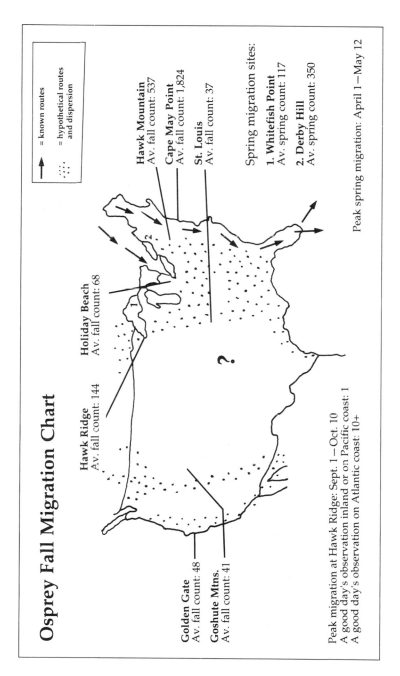

Legend:
→ = known routes
⋮ = hypothetical routes and dispersion

Hawk Ridge
Av. fall count: 144

Holiday Beach
Av. fall count: 68

Golden Gate
Av. fall count: 48

Goshute Mtns.
Av. fall count: 41

Hawk Mountain
Av. fall count: 537

Cape May Point
Av. fall count: 1,824

St. Louis
Av. fall count: 37

Spring migration sites:

1. Whitefish Point
Av. spring count: 117

2. Derby Hill
Av. spring count: 350

Peak migration at Hawk Ridge: Sept. 1 – Oct. 10
A good day's observation inland or on Pacific coast: 1
A good day's observation on Atlantic coast: 10+

Peak spring migration: April 1 – May 12

Peregrine Falcon
Falco peregrinus

OF ALL THE BIRDS IN THIS VOLUME, THE PEREGRINE FALCON IS CLEARLY THE rarest and most endangered. From the 1940s to the 1960s, the U.S. peregrine population east of the Mississippi was eliminated; at the end of that time not a single pair was known to breed there. Populations in other areas of North America were also dropping.

This decline is believed to be mainly a result of the increasing use of DDT and its derivatives as pesticides. Other small birds that the peregrine catches and eats picked up the poison from feeding on and near agricultural products. The ingested poison had the effect of making the egg shells of peregrines so soft as to crack under the weight of the incubating bird. By 1969, the peregrine was so rare that the United States listed it as an endangered species.

Since that time, heroic efforts have been made to reintroduce the peregrine falcon into areas where it otherwise no longer breeds and to bolster the populations at existing breeding areas. This immense job has been spearheaded and coordinated through The Peregrine Fund (see Appendix A).

Due to the dedicated work of many people associated with The Peregrine Fund, peregrine populations are slowly on the rise, and a few breeding pairs of peregrine falcons are now again living in areas where the birds were once common. The peregrine also seems tolerant of living in cities and on human-built structures, much to its benefit, and it can feed on city birds, such as pigeons.

Because there is such a strong interest in peregrines and because the birds may be more frequently seen these days in cities and elsewhere, we thought it important to include its behavior in this volume. Clearly, peregrines should be watched only from a great distance, with a scope or powerful binoculars. After all, we do not want to disturb the bird but to encourage it to nest wherever it can and wherever it chooses.

BEHAVIOR CALENDAR

	TERRITORY	COURTSHIP	NEST-BUILDING	BREEDING	PLUMAGE	SEASONAL MOVEMENT	FLOCK BEHAVIOR
JANUARY	▓	▓					
FEBRUARY	▓	▓	▓			▓	
MARCH	▓	▓	▓	▓	▓	▓	
APRIL	▓			▓	▓		
MAY	▓			▓	▓		
JUNE	▓			▓	▓		
JULY				▓	▓		
AUGUST						▓	
SEPTEMBER						▓	
OCTOBER							
NOVEMBER							
DECEMBER							

DISPLAY GUIDE

Visual Displays

Aerobatic-Flight

Male or Female   *W Sp Su*

Spectacular maneuverings in the air, generally near the nest, include high soaring, steep dives, figure eights, loop-the-loops, rolls, undulating flight, and chases. During the simultaneous and proximate display of two birds, talons or bills may touch. May be done by one or both members of a pair.

CALL: Creaking-call

CONTEXT: Done during courtship, which is mainly in late winter and spring, but also may be seen earlier in season. Seems to be part of courtship and, possibly, territorial advertisement. *See* Territory, Courtship

Slow-Flight

Male   *Sp*

Bird does a slow, bouncing flight with wings held high while flapped at their tips and tail lowered.

CALL: None

CONTEXT: Done by the male as he approaches the female and often just prior to copulation. At the end of the flight he rises up and then drops down to land. He may hold his wings hunched up for a few seconds after landing. *See* Courtship

Head-Low

Male or Female   *Sp Su F W*

The bird lowers its head in relation to the rest of its body, which may be in a vertical or horizontal posture. In appeasement behavior the bill is directed down or away from the other bird; in aggressive behavior the bill is pointed at the other bird. May be accompanied by Bowing.

CALL: Creaking-call or Wailing-call

CONTEXT: Most common display given by mated pair as they approach each other or greet each other on the ledge after being apart. Appeasement display common in the male at the approach of the female. *See* Courtship

Bowing

Male or female   *Sp Su F W*

Head and upper body move repeatedly up and down.

CALL: Creaking-call or Wailing-call

CONTEXT: May accompany Head-low display during interactions with mate. May also occur during aggressive encounters with other falcons or while watching prey. *See* Courtship

NOTE: There are other visual displays of the peregrine not included here because they are not as easily distinguished or commonly seen as the ones listed above. For more on visual displays see the articles listed in the bibliography.

Auditory Displays

Cacking-Call

Male or Female   *Sp Su F W*

A short, harsh, high-pitched sound repeated many times at a rate of two times per second. Male version higher pitched and less harsh than female version.

CONTEXT: Used during any disturbance near the nest, such as an intruding peregrine or other potentially dangerous bird and at the approach of any human. May be continued for several minutes. *See* Territory

Wailing-Call

Male or Female   *Sp Su F W*

A drawn-out, high-pitched sound, rising at the end; may be repeated.

CONTEXT: Given by adults when mate-feeding and bringing food to young in the nest; given by older nestlings when receiving food from parents; and given by adults during intense aggressive encounters with intruders. *See* Territory, Courtship, Breeding

Creaking-Call

Male or Female   *Sp Su F W*

A two-syllable call with the emphasis on the first syllable. Quite variable in character, but often described as sounding like a squeaky hinge. May be repeated several times.

CONTEXT: Given during courtship and during interactions between a mated pair near the nest. Most common during incubation and slightly before. *See* Courtship

NOTE: The calls described above are the most common ones loud enough to be heard from a distance. There are several other quiet calls given between mates and between parents and young.

BEHAVIOR DESCRIPTIONS

Territory

Type: Nesting, mating
Size: Within several hundred yards of nest
Main behavior: Soaring, Aerobatic-flight, Cacking-call, Wailing-call
Duration of defense: Throughout breeding and possibly longer

To what extent peregrine falcons are territorial is not clear. Where peregrines are common, their nests are fairly evenly spaced, but this may be more a result of natural avoidance than due to territorial aggression at borders.

The conspicuous aerial displays of peregrines, generally attributed to courtship, may in part advertise the presence of breeding pairs to other peregrines in the area, causing them to stay away.

Territorial interactions between peregrines occur mainly within several hundred yards of the nest. One form these encounters take is for the resident pair to give the Cacking-call and Wailing-call as they circle about the cliff, attempting to get higher than the intruder, who is usually silent. Once higher, they may swoop down upon the intruder. The intruder flips on its back and presents its talons to the diving bird. Actual hitting is not common, but occasionally the two may lock talons and fall down through the air together. They separate just before hitting the ground. This diving and chasing during territorial interactions is similar to courtship activity and can be hard to distinguish from it.

Direct aggressive behavior like this seems to be the exception rather than the rule. In the majority of cases a third bird just appears near the breeding pair, often circling higher than the pair, and then, with seemingly no interaction occurring, just drifts off.

Peregrines are aggressive toward other large birds nesting nearby, such as ravens. They will dive on them repeatedly whenever they fly near the peregrine's nest. Ravens defend themselves like hawks, by flipping over in flight and presenting their claws to the attacking bird. Both members of a pair may attack a single raven, all the while giving the Cacking-call. Peregrines may also be aggressive toward Accipiters, other falcons, buteos, ospreys, and turkey vultures.

Estimates of the home range of peregrines vary from about six to nineteen square miles. Nests may be as close as four hundred and fifty to five hundred yards apart, but are usually several miles apart. Besides defending the few hundred yards around the nest, the birds may defend other areas or perches farther away — used for hunting or plucking feathers off prey — throughout the year. In other cases, the hunting ranges of neighboring pairs may overlap.

Courtship

Main behavior: Aerobatic-flight, Creaking-call, mate-feeding
Duration: Late winter through early spring

Pairs of peregrines may occupy ledges at any time of year, but courtship and close association between mates occur primarily in late winter and early spring.

The first bird to occupy a nesting ledge may be a lone male or female that nested there before or is nesting for the first time. In some cases pairs return together to their nesting ledge.

A lone bird may advertise its presence to other passing peregrines by flying back and forth along the face of the cliff and giving the Creaking-call and by flying from ledge to ledge along the cliff and landing briefly at each spot, as if advertising the nesting potential of the cliff.

If a bird that has previously nested at a cliff arrives ahead of its former mate, it may engage in courtship with other peregrines that come along, but when the former mate shows up, this newly arriving bird chases off all other competitors.

Early courtship activities involve three behaviors: cooperative hunting, flight displays, and mate-feeding. Hunting is often done early in the morning. In the beginning stages of courtship the pair may just hunt in the same vicinity; later they may actually cooperate in hunting the same bird—one of them separating the bird from the flock and the other swooping at it.

After successful hunting, the pair may do exciting courtship flights. These involve both birds and are often preceded by the two soaring close together as they rise higher and higher. Following this, one or both birds start a series of aerobatic movements that may include steep dives, undulating flight, loop-the-loops, rolling over, and figure eights, all at incredible speeds. Chases may also be a part of these displays, with male or female chasing the other member of the pair, diving down on it from above with the other bird rolling over to face the diving bird with its talons at the moment the two come near. Occasionally, the two may lock talons or bills briefly in midair. The flights are usually accompanied by the Creaking-call and may be followed by additional soaring or by the two birds perching on a ledge. After several minutes the birds may repeat their displays. Display flights continue up to the time of egg-laying.

Mate-feeding, or the male's bringing food to the female, begins a week or two after courtship has started. In the early stages of cooperative hunting, the female may forcibly take food from the male by chasing him and grabbing it in her talons. She is the more dominant member of the pair at this time and can readily displace the male at any perch that she favors, making him fly off as she lands.

Later, the male may drop off food at a certain spot where the female then retrieves it. A more ritualized form of mate-feeding develops and continues for several months through the breeding season. In this, the male gives the Wailing-call as he flies toward the female with prey. She may also give the Wailing-call as she either flies out to meet him or waits for him on a perch. He lands near her, transfers the food to his beak, and then does some Bowing before she takes the food in her beak. The Wailing-call and Creaking-call may be heard during food transfer.

Variations on mate-feeding include: the male's just dropping food off for the female; passing it to her in midair, with his dropping it and her catching it; or her coming up underneath him, rolling over, and taking the food in mid-flight. Sometimes the male eats part of the food or does exaggerated feather plucking before giving the prey to her.

Once the female gets the food, she will take it to a special ledge to eat it. If some of the prey remains after she is finished, she will cache it in a crevice among the rocks. Courtship feeding continues throughout the breeding period.

Another element of courtship among peregrines is what has been termed ledge displays. These are actions performed by the birds on ledges or potential nest sites. They may be done alone or when the birds are together and include these displays: Head-low, Slow-flight, Bowing, Wailing-call, and the Creaking-call. They may also include three other actions: scraping (see Nest-building), touching bills, and copulation.

Copulation can become frequent in peregrines just prior to egg-laying and continuing into the egg-laying period. In copulation, the male lands on the female's back and presses his tail down to one side, while she crouches down, leans forward, and moves her tail off to the other side. Copulation lasts about ten seconds and may be repeated up to four or more times per hour.

When courtship occurs depends on the weather, the age of the pair, and whether the pair are residents or migrants. Many of the courtship behaviors (excluding copulation) described above have been seen in autumn, winter, and spring, but they most often occur in the months just prior to breeding. It also seems that pairs that have previously bred may start courtship earlier in later years.

As with most birds, there seems to be a population of undetermined size of unmated peregrines. These may be young birds that either are sexually immature or for some reason delay their breed-

ing activity. These lone birds may stay away from breeding peregrine territories. Sometimes they are referred to as "floaters." They
probably are the resource from which breeding peregrines draw
replacement mates when a member of a pair dies. In some cases,
peregrines have been known to get a replacement mate as soon as
a week after their mate dies.

Nest-Building

Placement: On the ledge of a cliff or building
Size: 12 inches in diameter, 1–2 inches deep
Materials: A depression scraped in the earth

Peregrines typically nest on the ledges of steep cliffs, often
about halfway up or down the cliff face. They prefer cliffs where
there are several ledges for nesting, roosting, eating, and food
transfer. Nesting cliffs often have a southern exposure, and birds
on the seacoast usually nest in protected coves. On the rocks just
below the aerie, you may see a whitewash of droppings or see
orange lichens, which seem prone to grow at that spot.

The nest is just a shallow depression scraped in soft substrate on
a cliff ledge. It is about twelve inches in diameter, one to two
inches deep, and usually contains no other material. Occasionally,
peregrines may nest in the tops of trees in a hollowed-out section
where the top of the tree has broken off. In some areas, where
there is no earth to scrape away, they may nest in a slight depression in the rock formed by erosion.

Nest building seems to start with the male's taking an interest in
certain ledges. Within sight of the female, he flies to a ledge, gives
the Creaking-call or Wailing-call, and scrapes with his feet as if
constructing the nest. He then flies off to another ledge and repeats the actions. At some point the female may join him on one of
these ledges and seem to take an interest in it.

When she starts to look for a ledge to nest on, she visits many of
them by herself, scraping and turning around on them as if to test
them for fitness and size. After choosing one ledge she completes

the nest in a day or two by scraping it out of the earth. She may use her bill as well as her feet in construction.

Some researchers think that nest-scraping is a display that helps to integrate the pair. Scrapes are made by both male and female at various ledges throughout the courtship period. The scrape finally used is often not in one of these early spots.

The peregrine typically changes its nest site on the same cliff from year to year. There is one record of change occurring ten years in a row with no repeated uses. In most cases, a previously used nest site is used again in several years. If the first clutch is destroyed then the birds also switch the site for a second clutch, sometimes to the one used the previous year.

Male and female roost at night in separate spots along the cliff, sometimes facing the cliff face as they sleep.

Breeding

Eggs: 3 or 4. Creamy or pink base covered over with rich brown blotches often concentrated at one end
Incubation: 28–33 days, mostly by female
Nestling phase: 4½–6 weeks
Fledgling phase: At least 6 weeks and usually several weeks more
Broods: 1

Egg-Laying and Incubation

Eggs are laid one every two to three days and a completed clutch averages three or four eggs. The female may lay the eggs at any time of day but does so most often in the early morning. Most laying occurs in late March and early April. If the clutch is destroyed anytime up to the tenth day of incubation, then the pair may produce a replacement clutch within about three weeks.

Incubation begins after the third or fourth egg is laid. Before that, one of the pair, usually the female, stands by the nest, facing out and guarding the eggs during the day and probably sitting over them at night. All nighttime and most daytime incubation is then done by the female.

During incubation the male does almost all food gathering for the female, giving her the food in midair or at a nearby perch. The female eats the food away from the nest and at this time the male usually goes to the nest and incubates. Occasionally, the female will leave the nest to preen, defecate, or hunt, in which case the male again may come to the nest and incubate.

During incubation, peregrines are wary of human presence and will usually take flight if you come within sight of the nest—even if you are still four to five hundred yards away. When disturbed, the birds may fly back and forth in front of the nest area giving the Cacking-call until you leave. Occasionally, the birds will stay put even when you are within a hundred yards of the nest; tolerance of human presence varies among individuals. Although human presence does not usually make peregrines abandon clutches during incubation, extreme care should be taken; you should remain as far as possible from the nest and use telescopes to observe, so as not to disturb the birds.

Incubation lasts from twenty-eight to thirty-three days, the length of time depending on a variety of factors, such as the weather, size of the eggs, and behavior patterns of the individual birds.

Nestling Phase

Although the eggs are each laid two to three days apart, they all generally hatch within a two-day period since true incubation does not begin until the last eggs are laid. An exception is in the far North; here incubation may start with the first egg, resulting in the young hatching over a period of a week.

For the first eight to twelve days after hatching the female broods the young almost continuously. After this time, the female rapidly tapers off her brooding during the day and may spend her time on a nearby perch or doing some hunting. However, she continues to brood the young at night until they are about three weeks old. The attending parent may also shield the young from the sun by spreading out its wings.

The young when first born are covered with white down and are quite conspicuous on the nest as they huddle together. For the first few days they mostly sleep, can hardly see, and cannot move out of the nest scrape. After the first week they can see and are more active on the nest; they respond to the parents by crawling over to them when they arrive. After about three weeks their wing and tail feathers begin to grow and the birds stretch and flap their wings more often. After four weeks their tail and wing feathers are more fully developed and their body feathers grow. Along with this feather growth comes more preening. At this age, the young move off the scrape and are so aggressive that the parents just leave the food off at the cliff and do not remain.

Throughout this period the male has continued to bring food to the female, who then feeds it to the young. Prey is usually delivered to the female plucked of feathers and without the head, which is probably eaten by the male. The female rips off pieces to feed the young. The young gradually develop the ability to eat on their own and pull bits of flesh off prey items. After the young are three weeks old, the female is more likely to leave the young and also to hunt. After four weeks the young can feed on their own when food is brought to the nest. They now can give a Wailing-call similar to that of the adult and do so when food is brought. It is loud and can be heard at a considerable distance from the nest. Prey items are brought to the nest four to eight times per day, often most frequently in the early morning and evening.

The male, at times, may cache a considerable amount of food in certain ledges of the cliff. This reserve food can counterbalance changes in the availability of prey due to weather conditions such as heavy rain.

In the later stages of the nestling phase the young may disperse to adjacent cliff ledges. It is important not to disturb the young and the nest at this stage of the breeding cycle, for you may cause them to fly prematurely and they may not be able to fly sufficiently well to reach another perch.

The nestling phase lasts from four and a half to six weeks.

Fledgling Phase

The males, which are smaller and develop faster, are usually the first to take flight. Once the young have left the nest, they roost in the general vicinity for a week or so and are fed by the parents. At first the food is still given to them in pieces and plucked of feathers, but later whole prey is transferred. Adult male and female may hunt together again and food brought to the young may be transferred in midair, with the young flying up underneath the adult and rolling over to get the prey item.

Increasingly, the young take flights off the cliffs and may engage each other in chases and aerial maneuvers similar to those of courtship in adults. Their ability to catch prey is still limited; they may chase after birds without catching them or catch insects in midair. Over the weeks the family may move off to other cliff areas for a while. Gradually the young and parents separate, although feeding of young by parents has been seen as late as September. How and when the family actually separates is not known. The period during which the fledged young are dependent on the parents for food is at least six weeks.

In some cases in the North, family groups stay together up until migration, at which time they probably break apart. In other cases the young may be on their own for several weeks before migration.

Plumage

DISTINGUISHING THE SEXES There is no way to distinguish between the sexes on the basis of plumage; however, the female is larger than the male. In terms of behavior, the female does almost all of the incubation and brooding, and during this time the male brings all food to her.

DISTINGUISHING IMMATURES FROM ADULTS Adults are slate gray on the back and have faint barring mostly on the lower breast; immatures are dark brown on the back with buff edges on their wing feathers and dark streaking over the entire breast. This immature plumage is kept until the bird's first spring, when it starts to molt and acquire adult plumage.

MOLTS Peregrines have one complete molt per year. For most peregrines this molt starts in late spring and continues into fall, lasting from four to six months. For birds breeding in the Arctic, the molt starts in midsummer, stops during migration, and resumes on the wintering ground. In most cases, male molting seems to start and end later than female molting.

Seasonal Movement

In North America, the peregrine falcon is migratory, with the breeding populations of Alaska and Canada flying south in September and October and north again in March and April. Most migration seems to take place along the coasts but some also occurs inland. The birds may winter along the Atlantic, Gulf, and Pacific coasts, but most seem to fly to Central and South America for the winter. A few populations on the West Coast and in the Rocky Mountains may be year-round residents.

RECOGNITION DURING MIGRATION A large, powerful hawk with long tapered wings. The bird tends to flap its wings regularly and constantly unless soaring. With its tremendous speed and grace it seems to slice through the air effortlessly. A close-up view may reveal its black mustache, visible down either side of its face.

BEHAVIOR DURING MIGRATION Flies singly and often continuously, usually at fairly low altitudes and often along the coast. It may remain at certain spots along its route for a while and chase after small migrating birds that it feeds on.

For more information on hawk-watching, see Appendix B, Hawk-Watching.

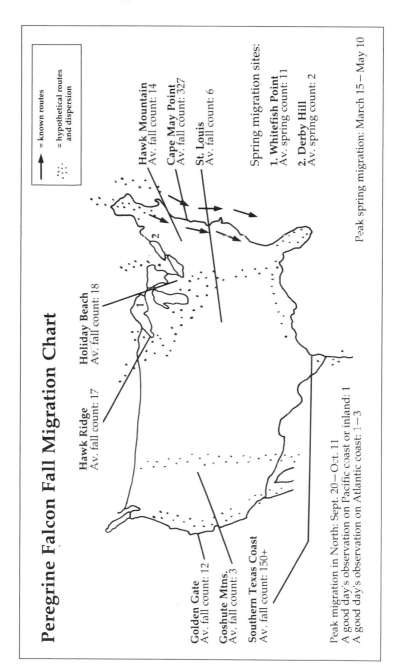

Peregrine Falcon Fall Migration Chart

→ = known routes

⋮ = hypothetical routes and dispersion

Hawk Mountain
Av. fall count: 14

Cape May Point
Av. fall count: 327

St. Louis
Av. fall count: 6

Holiday Beach
Av. fall count: 18

Hawk Ridge
Av. fall count: 17

Golden Gate
Av. fall count: 12

Goshute Mtns.
Av. fall count: 3

Southern Texas Coast
Av. fall count: 150+

Spring migration sites:

1. Whitefish Point
Av. spring count: 11

2. Derby Hill
Av. spring count: 2

Peak spring migration: March 15 – May 10

Peak migration in North: Sept. 20 – Oct. 11
A good day's observation on Pacific coast or inland: 1
A good day's observation on Atlantic coast: 1–3

Bob Hines

Northern Bobwhite
Colinus virginianus

FOR THE BEHAVIOR-WATCHER, THE BOBWHITE IS A REAL CHALLENGE SINCE these birds are so wary of human presence. Their most conspicuous behavior is their calls: the Bobwhite-call in spring and summer, which is given primarily by unmated males, and the Separation-call in fall and winter, which is given by covey members at dusk and dawn.

As you get near the birds it is likely that they will see you first and go into alarm behavior. This may involve the Psieu-call and distraction display in summer when the parents are attending their brood, and the Psieu-call or Toilick-call in fall and in winter if you happen to come upon a covey.

Another outstanding behavior of bobwhites is their roosting formation in the fall and winter coveys. Here the birds arrange themselves in a tight circle, side by side and tails pointing inward, as a means of conserving heat. In light snow, tracking the birds may lead you to one of these roosting locations where you may even see a ring of droppings left behind.

BEHAVIOR CALENDAR

	TERRITORY	COURTSHIP	NEST-BUILDING	BREEDING	PLUMAGE	SEASONAL MOVEMENT	FLOCK BEHAVIOR
JANUARY							■
FEBRUARY					■		■
MARCH					■		■
APRIL	■	■			■		
MAY	■	■	■	■	■		
JUNE	■	■		■			
JULY	■			■			
AUGUST				■	■		
SEPTEMBER					■	■	
OCTOBER					■	■	
NOVEMBER					■		■
DECEMBER							■

DISPLAY GUIDE

Visual Displays

Frontal-Display

Male   *Sp Su*

The bird's body is horizontal, with tail fanned, body feathers fluffed, and wings spread and rotated forward.

CALL: None

CONTEXT: Stance of a male when he and his mate are challenged by another male. May also be done between two unmated males, in which case it may determine which is dominant. Occasionally done toward the female at the start of courtship. *See* Courtship

Lateral-Display

Male   *Sp Su*

Male walks slowly in front of female with head low and with tail spread and tilted toward female.

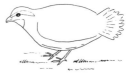

CALL: None

CONTEXT: Given during the breeding season. May be followed by copulation. *See* Courtship

Tidbitting

Male   *Sp Su*

Bird arches forward, fluffs body feathers, and pecks at or picks up food items. Tail may be fanned.

CALL: Tu-tu-call

CONTEXT: Done during the breeding season. Usually results in the female's coming over and taking bits of food from the male, as in mate-feeding in smaller birds. Tidbitting as a courtship display involves both the actions and associated call. At other times of year, just the call is given and this directs the attention of offspring or other covey members to a food item. *See* Courtship, Tu-tu-call

Wing-Quiver
Female   *Sp Su*

The female rapidly raises and lowers her wings as she walks toward the male.

CALL: None

CONTEXT: Done by female to male during early stages of courtship. Once birds are paired it is no longer done. *See* Courtship

Auditory Displays

Bobwhite-Call
Male   *Sp Su*

A two- or three-part whistle, sounding much like the written words "bob-white," or "ah-bob-white." The syllables are distinct and the last note is a rapid upward slur. Variations on this call involve one of the syllables being repeated several times. Generally four to eight calls are given per minute.

CONTEXT: A common call, given most frequently and loudly by unmated males in late spring and early summer as they announce their presence on their whistling territories. Duets between neighboring males may occur. Mated males give the call much less frequently and more softly. *See* Territory, Courtship, Breeding

Separation-Call

Male or Female   *W Sp Su F*

Varies from a loudly whistled "koilee" to a moderate "hoypoo" to a quiet "hoy." May be done several times in quick succession.

CONTEXT: A common call. Loud version given by members of a covey just before roosting at night and just after waking at dawn; may be answered by other nearby coveys. Softer versions given by a covey member that is temporarily separated from the others. Also given between mates that are separated or by females who have lost a mate. *See* Flock Behavior, Courtship

Caterwaul-Call

Male or Female   *W Sp Su F*

A rasping call given as a set of short, loud, separate syllables. Generally three to five syllables. Example: "Hao poo weih."

CONTEXT: Given by males mostly just before and during the breeding season, but can be heard at other times of the year. Occurs during aggressive encounters between rival males when they are near a female. Occasionally given by females when faced with a rival female. *See* Territory, Courtship

Psieu-Call

Male or Female   *Su*

A thin, cheeping sound given by chicks and adults. Sounds like baby chickens peeping.

CONTEXT: Given when a ground predator disturbs the brood or the nest. May also accompany the distraction display of the adults. *See* Breeding

Toilick-Call

Male or Female   *Sp Su F*

A regular, monotonous call with a short, final syllable that is repeated a variable number of times. Example: "toilick, ick, ick, ick."

CONTEXT: Given by parent or young in the presence of predators. Continued until the predator leaves. *See* Breeding

Tu-Tu-Call

Male or Female   *W Sp Su F*

A soft, short, rapidly repeated call.

CONTEXT: During the early breeding season, this call is given by the male when he finds food and does the Tidbitting display, inciting the female to come over and feed. At other times it is given by males or females when they discover good sources of food; attracts other birds to the spot. *See* Courtship, Breeding

OTHER VOCALIZATIONS: Bobwhites are believed to have one of the most complex vocal repertoires of all quail. The calls listed above are the ones most commonly heard and the ones most important to understanding the general behavior of bobwhites. However, as many as twenty-four distinct vocalizations have been attributed to these birds.

BEHAVIOR DESCRIPTIONS

Territory

Mated Males
Type: Mating, nesting, feeding
Size: About 1 acre
Main behavior: Frontal-display, chase
Duration of defense: Early part of breeding season

In mid-spring, mated pairs of bobwhites leave the winter coveys (see Flock Behavior) and move to nearby areas where they will nest. The males may give the Bobwhite-call but usually not frequently or loudly. Rather, they are more quiet than unmated males and spend most of their time going about their breeding behavior. They do, however, tend to remain in a very restricted area of about one acre. Other nesting males may be as nearby as fifty feet. If another male approaches, the male will actively defend his mate with the Frontal-display and possibly chase the other male away.

Unmated Males
Type: Singing
Size: Unknown
Main behavior: Bobwhite-call, chases
Duration: Spring to midsummer

In spring and summer most of the Bobwhite-calls are given by unmated males. The territories of these males are often located around the areas where mated pairs are breeding. They call loudly and often and seem to defend the area from which they call against other unmated males, chasing them out. This behavior continues into midsummer. After then, defense of the area and Bobwhite-calls decrease.

Since, in general, there are more males than females in the population, it is believed that these unmated males may stay at the edge of nesting pairs' areas in the hope of replacing the mated male if he dies.

Courtship

Main behavior: Frontal-display, Lateral-display, Tidbitting
Duration: Throughout breeding

Bobwhite courtship is rarely seen in the wild since the birds are wary of human presence. Nevertheless, several aspects of their courtship have been studied in captive birds and we summarize them here so that you can identify courtship if you are lucky enough to witness it.

When and how bobwhites pair is not clearly understood. Some researchers feel it occurs in early spring around the time the first Bobwhite-calls are heard; others feel it may occur in winter within the covey.

Several displays believed to be associated with courtship occur mostly at the beginning of the breeding period. One of these is the Frontal-display, a primarily aggressive action that is used in a variety of circumstances, such as: when a mated male is confronted by a strange male or female; when two unmated males are competing for dominance; or when an unmated male meets a new female. Once a male is mated, he no longer directs this display at his mate.

Another early courtship behavior is the Lateral-display. This is done by a male toward his mate and may precede copulation. Also, at the beginning of courtship the female may do Wing-quiver. The exact function of this display is not known, but it may both attract a male's attention and inhibit any aggressive actions he might at first be liable to take.

One display of bobwhites that occurs throughout the breeding period is called Tidbitting. In this, the male, after discovering a morsel of food, gives the Tu-tu-call, bends down to peck at or pick up the food, and fluffs out his body feathers. This usually attracts his mate, who picks up the food or takes it directly from his bill. This behavior is called mate-feeding in other species.

Once two bobwhites are paired they spend all of their time together, feeding in the morning and afternoon, loafing at noon, and roosting together at night.

Nest-Building

Placement: On the ground
Size: 4 – 5 inches in diameter
Materials: Grasses, moss, pine needles

Nest-building by bobwhites in the wild is rarely seen since the birds are so wary. The nest is usually built under some slight protection of sparse vegetation and within fifty feet of an opening or clearing, such as a field.

At first the male or female pecks and scrapes a slight depression in the ground. The birds then draw in material from around the edge of the nest, such as grasses or weed stems. Sometimes they gather material from a few feet away by facing away from the nest, picking up material in their bills, and tossing it over their backs and toward the nest. This action may then be repeated closer to the nest, with the birds finally putting the material in the nest.

The whole nest is completed in two to three hours of work and may be partially covered over with grasses at the end. As do many other birds, bobwhites occasionally start nests and then abandon them, possibly in favor of a better site.

Breeding

Eggs: Average 12–14. White
Incubation: 23 days, by female in most cases
Nestling phase: None
Fledgling phase: 2–3 months
Broods: 1

Egg-Laying and Incubation

Egg-laying usually starts several days after the nest is completed. One egg is laid every one to one and a half days until the clutch is complete. The nest is not attended during the egg-laying period except during actual laying. Occasionally "dump nesting" occurs, whereby one female will lay eggs in another's nest; this results in a clutch of more than twelve to fourteen eggs. Also, occasionally other species, such as pheasants or domestic chickens, lay their eggs in bobwhite nests. Why this occurs is not known. All these eggs are usually accepted and incubated by the nesting female.

Incubation starts after the laying of the last egg and continues for twenty-three days. The female starts the incubation, but if something happens to her, the male may take over. Rarely do the pair share these duties. The incubating bird usually leaves the nest in the afternoon, sometimes also in the morning, each time joining its mate; they feed together. When one bird is incubating, the other stays away from the nest. The male may do some soft Bobwhite-calls during this time.

Many nests are lost to predation and the birds will then try to renest throughout the summer until they are successful. These later clutches often contain fewer eggs.

If the bird on the nest is flushed by a potential predator, it may do a distraction display, fluttering along the ground, dragging its wings, and giving the Psieu-call. This may have the effect of leading the predator away from the nest. If it is successful in doing this, the bird stops the display and secretively returns to the nest.

If the predator does not leave the nest site when the adult does the distraction display, then the bird gives the alarm call and continues it until all danger is gone.

Nestling Phase

The eggs hatch within an hour of one another. The young are able to leave the nest within an hour or two of hatching. If they hatch late in the afternoon, the parents remove the egg shells from the nest and the young and parents stay there through the night. If they hatch earlier in the day, then the parents lead the young away from the nest as soon as their feathers are dry.

Since the young are able to leave the nest right after hatching there is no real nestling phase.

Fledgling Phase

The family stays together all of the time, with either parent frequently brooding the young during the first two weeks to help keep them warm until their own feathers are well developed. This is especially true in the first week, when they are brooded most of the time, with brief periods of feeding in between.

While feeding, the family remains as a tight flock. When either parent finds a particularly good food it gives the Tu-tu-call and this causes the young to approach immediately and peck at it. Bobwhites eat a lot of insects, seeds, and fruits.

When threatened by a possible predator, the whole family gives the Psieu-call and runs about; the young hide while the adults may do the distraction display described previously under Egg-Laying and Incubation. If a predator persists the parents may give the alarm note — the Toilick-call — until it leaves. Then the family reunites and continues feeding.

At two weeks the young can already fly short distances and within a few more weeks can fly quite well. When the young are older and disperse farther in the face of danger, the Separation-call is given when danger is passed. This causes the family to regroup. This same call is used to bring together the members of a covey in fall and winter, after they have been separated. At about fifteen weeks the young are the same size as the adults.

Multiple family groups may join together in fall to form large coveys. *See* Flock Behavior

Plumage

DISTINGUISHING THE SEXES The major distinction between male and female is on the head and neck. The male has a bright white throat patch and a white streak over the eye, while in the female these are both buff colored.

DISTINGUISHING JUVENILES FROM ADULTS Up to about three months of age, juvenile bobwhites of both sexes are similar in appearance to adult females, though smaller and with duller coloration.

MOLTS Bobwhites undergo two molts per year. Starting in early spring or as early as February they have a partial molt of their head and neck feathers. This molt usually ends in May or June.

In August or September the birds undergo a complete molt of all feathers. This is completed by October or November. Neither molt involves a change of feather color.

Seasonal Movement

After the young have matured in late summer there is a general movement of bobwhites to good feeding areas for the winter. Groups of up to thirty or more birds, composed of families, un-mated adults, and unsuccessful pairs, may gather in an area to feed and roost together. There is some shifting of birds from group to group in September and October, but by November coveys have settled into their winter ranges and remain fairly stable in membership. This period of movement is sometimes called the "fall shuffle," and although most birds do not move more than about a half mile or so, a few may move as far as ten miles.

Flock Behavior

Bobwhites are well known for forming social groups in winter called "coveys." A covey is a group of birds that remain together through winter, feeding and roosting together in a fairly fixed range.

Coveys usually contain about twelve to sixteen birds. Some members are paired adults, others single, and others young of that year. One of the striking features of a covey is its roosting behavior. Each night the covey gathers together in a protected spot on the ground and forms a tight circle, each bird side by side to two others with all tails pointing inward. They then lift their wings slightly to form a continuous cover and remain that way till dawn. This helps the birds preserve heat at night.

Coveys can change in number of members in two ways. One is to have members die and the other is to have members from nearby coveys join them. When a covey is reduced in number to fewer than seven or eight birds it is no longer able to form a circle that holds sufficient heat and the birds may disperse and join other nearby coveys. When a covey comes to include more than about sixteen members it may roost in two circles at night.

Each covey remains on a range of from a few to thirty to forty acres, depending on how abundant food is. Ranges of neighboring coveys may overlap, for these areas are not defended.

At dawn the covey leaves the roost site together. Just before leaving, one or more members may give the Separation-call. This may in turn be answered by other neighboring coveys. The function of this exchange is not known but it may help coveys keep track of one another's movements and whereabouts. The covey feeds together in the morning, then at midday the birds rest in the sunlight, staying warm and digesting their food. In very cold weather they may form a circle at this time also. In mid-afternoon they again feed, and at sundown they gather at their roost site. Before forming their circle roost for the night they may give the Separation-call.

In late winter, paired birds within the covey start to feed alone during the day and rejoin the covey at night. In March and April these pairs finally leave the covey and move to their breeding area. This leads to the breakup of the covey.

Ring-Necked Pheasant
Phasianus colchicus

ALTHOUGH RING-NECKED PHEASANTS ARE FAIRLY SECRETIVE, A GREAT deal of their behavior can be observed. For example, one winter our neighbor complained that a pheasant was waking him up each morning just before dawn. The bird was standing on a big rock in his yard, giving a loud "skwagock" call, and then fluttering his wings. This is Crowing, which is done by the male and is a sign that courtship and territory formation are about to begin.

If you live in a suburban or rural area, you can attract pheasants by sprinkling cracked corn on the ground. One spring morning as we watched a pair of pheasants at our feeder, the male suddenly tilted sideways in front of the female, holding the wing nearest her spread down toward the ground and his tail turned toward her and fanned. He strutted in a measured way for a few steps and then returned to his normal position. The female continued to feed. This is a spectacular display that helps make sense out of all the brilliantly colored feathers of the male. It is the Lateral-display, one of the main features of pheasant courtship.

Later in summer, you may see the female and her brood of chicks crossing a road or wandering through fields. And in fall or winter you may see the groups of females that tend to feed and, generally, stay together during these seasons.

This is the way behavior-watching occurs with pheasants in the wild. You will rarely be able to see a complete sequence, since the birds are so wary, but you can put into context the bits of behavior you see in the long run. Over time, more and more pieces of the puzzle get added and you will get a sense of the bird's whole life.

BEHAVIOR CALENDAR

	TERRITORY	COURTSHIP	NEST-BUILDING	BREEDING	PLUMAGE	SEASONAL MOVEMENT	FLOCK BEHAVIOR
JANUARY							▓
FEBRUARY							
MARCH	▓					▓	
APRIL	▓	▓				▓	
MAY	▓	▓	▓	▓			
JUNE				▓			
JULY				▓	▓		
AUGUST				▓	▓		
SEPTEMBER					▓		
OCTOBER					▓	▓	
NOVEMBER							▓
DECEMBER							▓

DISPLAY GUIDE

Visual Displays

Lateral-Display
Male or Female   *Sp Su*

Bird turns sideways to recipient of display; tail is spread; the wing facing the recipient is drooped; the pinnae on the head are raised, and the wattles are swollen. The head is held tucked in close to the body or lifted up high, depending on the circumstances. Bird walks or struts in a shallow arc in front of recipient.

CALL: None

CONTEXT: When done as courtship by males to females the head is tucked in close to the body. When done as a territorial display to other males the head is held high. Occasionally done between females. *See* Territory, Courtship

Run-Threat
Male   *Sp Su*

With tail up, head up, pinnae raised, and wattles swollen the male runs, or, occasionally, walks, after another male.

CALL: None

CONTEXT: Given by male when asserting his dominance over another male during territorial disputes. Intruding male has tail low and wattles unswollen. *See* Territory

Tidbitting
Male   *Sp Su*

Male faces female, fluffs body feathers, and jerks head down toward a bit of food.

CALL: Tidbitting-call

CONTEXT: Done during courtship by male toward female. She usually comes over and then takes food. *See* Courtship

OTHER DISPLAYS: Two other displays are done by the female but are rarely seen. They are listed here so you can be aware of them. Both occur when the female is near the male and during what appears to be courtship. They are: 1) a short hop, and 2) a movement in which the female stretches her neck up and slightly opens her wings.

Auditory Displays

Crowing
Male   *Sp Su F W*

A loud two-part call with the emphasis on the second syllable. Preceded by several flaps of the wings and followed by an audible wing-whirr. Sounds like "skwagock." Given as the bird stands in an erect posture.

CONTEXT: Given by the male during territory formation. Usually heard most within a half hour before sunrise. Starts occurring regularly in late March and increases in frequency into May, then decreases in frequency into the summer months. May occasionally be heard in other seasons.

Cackle-Call
Male   *Sp Su F W*

A three-syllable call repeated several times. The call gets quieter at the end of the series. "Tucketuck, tucketuck, tucketuck, tucketuck."

CONTEXT: This may be given during flight or while on the ground. Given when bird is disturbed and when it takes flight. Sometimes given in the evening. May stimulate another male also to give the call. *See* Flock Behavior

Tucket-Call
Male   *Sp Su F W*
A two-syllable call, similar to the Cackle-call but not trailing off at the end and not given in flight.
CONTEXT: Possibly an alarm call, for it is given by males that are startled. Not heard as much during the breeding season.

Tidbitting-Call
Male or Female   *Sp Su*
A distinct series of clucking sounds: "kut-kut-kut."
CONTEXT: Given by the male when doing Tidbitting in front of the female. Given by female when with fledglings. *See* Tidbitting, Courtship, Breeding

Kia-Kia-Call
Female   *Sp Su*
A two-part, harsh call given by the female.
CONTEXT: A call given by female in response to the Crowing of the male. The exchange of calls between mated pair may help the two keep in contact. *See* Courtship

Krah-Call
Male   *Sp*
A hoarse, drawn-out call. Wattles may be swollen.

CONTEXT: Given in aggressive interactions between males. *See* Territory

BEHAVIOR DESCRIPTIONS

Territory

Type: Mating, feeding
Size: 3 – 9 acres
Main behavior: Crowing, Lateral-display, Run-threat, chases
.*Duration of defense: From early April until incubation phase*

A general dispersal of pheasants away from wintering areas and out over favorable breeding habitats occurs in late winter and early spring. This dispersal is in part caused by the increasing aggression among males.

At this same time, Crowing becomes a conspicuous feature of male pheasant behavior. It most likely functions to announce territory ownership and to attract a female. Crowing occurs in the early morning and again in the late afternoon. Males may Crow from the same spot for several days in a row, usually from a conspicuous location out in the open, such as on top of a rock or log. Individual males have distinct crowing calls, so with practice you can come to recognize individuals.

In some cases when one male crows neighboring males immediately give a series of high-pitched sounds starting soon after the crowing begins and drowning part of it out. This may function as some sort of continued communication between neighboring males, or it may be a way to make a neighbor's crowing less effective.

Territories are established in early April. At this time males that may have tolerated other males in their area during winter chase out all other males when they see them. Crowing at this time increases in frequency.

Older males return to the territories they occupied in previous years. The best territories are the first to be established. These are occupied by dominant males near areas where they winter. When an older male dies, the neighbors take over part of his territory.

Territories are about three to nine acres in size and contain both open ground with available food and areas of cover. The size depends on the quality of the habitat and the population pressure. As males get older their territories tend to expand, especially as neighboring territory holders die.

First-year males that try to establish territories are usually forced by older, more dominant males into marginal habitats. Most males do not establish territories until their second year; instead, they remain nonterritorial floaters, moving widely about. These floater males are tolerated in the territories of other males *if* they do not display and remain quiet. Thus in early spring when most males are intolerant of intruders, you may still see two males near each other; one is probably a first-year male.

In other cases when two males come near each other in spring, there will almost always be an aggressive interaction. These interactions between males are, generally, of three different types. In one, the two males will stand facing each other or walk along side by side. Their neck feathers rise and their wattles swell and they give the Krah-call. One of the birds may then give way, but if one does not, fighting may follow.

Another type of aggression is the Run-threat, in which the territory owner runs toward or after the intruder with head and tail high and wattles swollen.

One male may also do the Lateral-display to another male. However, unlike the display when done to a female, in this form the male stays still with his head held high, and his drooped wing is not as fully spread. This may be followed by the displaying bird's doing the Run-threat display as it chases the intruder.

In fighting, two males crouch down, close together and facing each other. They may peck at each other and, intermittently, one or both may jump up into the air. Fights may be prolonged, lasting an hour or more, but usually last about fifteen minutes. After fighting the two may walk parallel to each other, each one seemingly trying to get ahead of the other. As with a similar display in mockingbirds, this behavior may help to settle territorial borders.

Territorial skirmishes often occur when a male is following his harem and the females wander onto another male's territory.

Males stop defending any territory when all of their females have left to start incubation. Their territories may then be claimed by other males who still have females.

Courtship

Main behavior: Chases, Lateral-display, Tidbitting
Duration: From April until incubation phase

Pheasants have a mating system in which groups of females (harems) are claimed by single males. Groups of females that have gathered over the winter move to good feeding areas in late winter and early spring. Membership in these groups may change slightly as some leave the group and others join it, but a harem generally becomes stable in membership by May.

Females form bonds with the male whose territory encompasses their chosen feeding area. The male's bonds with each female are gradually established through courtship displays. Females are believed to be monogamous and tend to mate with the same male in successive years.

Not all male pheasants acquire harems. About 30 percent of the males in an area may have no territories at all, and, consequently, no mates. Another 25 percent may have a territory but only one or no females. Thus, only about 45 percent of males tend to have more than one female.

Most courtship occurs in the morning when the harem leaves the protection of cover to feed in the open. Males tend to display to females that are temporarily separated from the group. In fact, if that female rejoins the group, the male will stop displaying.

Early courtship may involve the male's taking short runs after the female over a distance of one hundred yards or more. Later, courtship involves two displays of the male: the Lateral-display and Tidbitting.

The male usually does the Lateral-display while strutting in an arc in front of the female. If she does not move, he holds the display in front of her; if she moves, he follows her and repeats the display. The display may be given ten or more times in succession.

Several actions of the female may stimulate the male's Lateral-display. These include a hopping movement and an action in which she stretches her neck up near him.

In Tidbitting, the male gives the Tidbitting-call as he fluffs out his body feathers, droops his wings, and makes pecking motions at a food item. This usually causes the female to come over and eat the food.

Late April and early May are when most copulations occur. Copulation is preceded either by the male's doing a series of rapid Lateral-displays, the female's squatting, and the male's mounting, or by the female's squatting with no displays by the male and the male's mounting. This latter behavior can be more common later in the breeding season.

Floater males may try to attack females in harems and force-mate with them, especially when the territorial males are otherwise engaged in a dispute. These nonterritorial males do not Crow, or do any other displays to the female. Instead, a male will simply run after a female, grab her neck feathers in his bill, and attempt to mount.

Harem size decreases in late May and June as the females leave to start incubation.

Nest-Building

Placement: On the ground
Size: 8–12 inches in diameter, shallow
Materials: Grasses, leaves, weed stalks

To make the nest the female scratches a shallow hole in the ground and lines it with grasses and leaves. The nest may be in the woods or a meadow. Hay and weed fields are a preferred habitat.

The female generally nests just outside her normal home range, as established during the time she was in the harem, and often just outside the territory of her mate.

Breeding

Eggs: Usually 10–12 eggs. Brownish olive in color
Incubation: 24 days, by female only
Nestling phase: None
Fledgling phase: 10–11 weeks
Broods: 1

Egg-Laying and Incubation

Egg-laying starts in April and reaches a peak in May. The female leaves the harem to lay each egg and then returns to join the other females. One egg is laid about every one and a half to two days. When laying the first egg, the female spends about one to two hours on the nest. With later layings she may spend up to six hours on the nest, but none of this time is real incubation. Younger females tend to lay fewer eggs than older females. Anywhere from nine to seventeen eggs may be in a clutch.

Pheasants have been known to lay their eggs in the nests of other bird species, such as mallards, woodcocks, ruffed grouse, and bobwhites. Why this occurs is not known.

Incubation starts with the laying of the last egg. As of this time the female no longer joins the harem. Incubating females tend to leave their nest only once a day and that is often in the late afternoon for about an hour. During that time they feed and preen. Incubation lasts for twenty-four days.

During the first week of incubation the female is very likely to desert her nest and eggs if disturbed by a potential predator, such as a human walking too close. If disturbed in the later stages of incubation she is more likely to return to the nest. If she deserts her nest, she may start renesting in as soon as ten days. Nests found in mid- or late summer are not second broods but renesting attempts after first nestings have failed.

Nestling Phase

Most hatching of young pheasants occurs in June. Hatching in a single nest may occur over a period of twelve hours. The young leave the nest immediately, so there is no nestling phase.

Fledgling Phase

The young are brooded by the female for the first day, usually in the immediate area of the nest. Over the next few days the female leads the young farther from the nest but usually stays within a five-to-ten-acre area around the nest. After that, the female and her brood may wander more widely in search of good cover and food. They tend to feed just in the early morning and late afternoon and rest under cover for the remaining time. The female broods the young every night.

If disturbed, the female may do a distraction display in an attempt to lead you or a predator away from the young. This may look like an injured and intermittent flight.

In general, the male does not join the female and young, but there are records of males being seen with a female and her brood when the young are over six weeks old. At about ten to eleven weeks the young disperse and feed on their own.

The young can be distinguished by sex when about five weeks old, the male having a bare patch on the side of his face whereas the female is feathered there. After a little over four months the young are in their first adult plumage.

Plumage

DISTINGUISHING THE SEXES The sexes of adults are easily distinguished by plumage. The female is slightly smaller and mottled brown on her sides, back, and shorter tail. The male has a dark head, white neck ring, a dark red-brown body, and a long tail. He also has two distinctive features on his head: red wattles like those of a chicken on the sides of his cheeks, which swell up and enlarge during certain displays, and feather tufts on his head, called pinnae.

DISTINGUISHING JUVENILES FROM ADULTS Juveniles' colors are similar to those of adult females, but they have smaller bodies and shorter tails than adult females. Young males are one quarter to one third heavier than young females.

MOLTS Ring-necked pheasants have one complete molt per year, starting in July or August and ending in September or October.

Seasonal Movement

There are two times during which pheasants tend to move or shift location: in October, when they move to wintering areas chosen for cover, and in March and April, when they disperse to breeding areas.

Flock Behavior

After the breeding season, pheasants gather into small flocks that feed and roost together throughout fall and winter. These groups are usually made up of all males or all females. Male groups are smaller, often including only two birds, as opposed to female groups, which average four or five birds. There may be some shifting of members between groups, especially among males.

In February males begin to interact more aggressively, and this leads to the dispersal of males over the area and a gradual breakup of their winter flocks.

Great Horned Owl
Bubo virginianus

WHETHER YOU HEAR THE GREAT HORNED OWL WHILE OUT CAMPING OR through an open bedroom window, all the wildness of the woods seems to accompany the sound. Its low muffled hoots, heard most in the early evening or pre-dawn hours, sound sedate and aloof and may be answered in an unhurried way by another owl far away. I remember once being startled by hearing a great horned owl call from right above me in a white pine tree at about ten o'clock at night. The sound, which from a distance is so soft, at this close range seemed to make my lungs vibrate.

The great horned owl is the largest of our common owls and is a predator on many large mammals as well as other birds. It has even been known to prey at night on ospreys and great blue herons. But despite this record of ferocity, it still must endure the constant harassment of crows during the day. Why crows mob great horned owls, as well as other owls and hawks, is not clearly understood. It does not make the owl leave the area more than temporarily, and it takes quite a bit of energy on the part of the crows.

Seeing or hearing mobbing behavior is one of the easiest ways to locate these owls. Just listen for the drawn-out caw that the crows give when mobbing and then rush to the spot and look at where they are diving. This will often lead to a glimpse of the owl as it takes a short flight to try to rid itself of the crows.

BEHAVIOR CALENDAR

	TERRITORY	COURTSHIP	NEST-BUILDING	BREEDING	PLUMAGE	SEASONAL MOVEMENT	FLOCK BEHAVIOR
JANUARY	■	■	■	■			
FEBRUARY	■	■		■			
MARCH	■			■			
APRIL	■			■			
MAY	■			■			
JUNE	■			■			
JULY	■			■	■		
AUGUST					■	■	
SEPTEMBER					■	■	
OCTOBER					■	■	
NOVEMBER	■						
DECEMBER	■	■					

DISPLAY GUIDE

Visual Displays

Wing-Spread
Male or Female   *Sp Su F W*

Wings are spread to the side and tilted forward, and body feathers are fluffed, making the bird look extremely large. Bill is gaped.

CALL: Hiss-call or Bill-snap

CONTEXT: Used during aggressive or defensive interactions with other owls or predators. *See* Territory, Courtship

Bowing
Male or Female   *W Sp Su*

Bird repeatedly bows its body forward and down.

CALL: None or various screams, barks, and whistles

CONTEXT: Given between mated pair during courtship and during greetings or when changing places at the nest. *See* Courtship, Breeding

Auditory Displays

Hoot
Male or Female   *Sp Su F W*

A series of four or five deliberate, deep, resonant hoots. Given in various rhythms by different individuals. Female hoots are shorter and higher-pitched than those of the male, even though she is a larger bird.

CONTEXT: Given in all seasons, but most commonly in fall and winter by the male as he advertises and defends his territory. May also be used during times when the owls are disturbed, such as by a possible predator near the nest. *See* Territory, Courtship

Hiss-Call
Male or Female   *Sp Su F W*
A hissing sound.

CONTEXT: Given during potentially hostile interactions. Often accompanied by the Wingspread display.

Bill-Snap
Male or Female   *Sp Su F W*
Bill is snapped shut rapidly and repeatedly, making an audible clicking sound that can be heard thirty to fifty yards away.

CONTEXT: Given during aggressive interactions. An owl may Bill-snap at crows that are mobbing it.

OTHER SOUNDS: These owls give such a wide variety of other sounds that it is impossible to describe them all or sort them out by their functions. Usually combinations of barks, screams, and whistles occur during interactions between the pair or during territorial aggression.

BEHAVIOR DESCRIPTIONS

Territory

Type: Nesting, feeding
Size: ⅓ – 2 square miles
Main behavior: Announcing, patrolling, calling, and possible fights at
common borders
Duration: From about a month before breeding until after breeding

Great horned owls are clearly territorial. In general, males begin to occupy breeding territories in November and start to hoot from various perches within the territory. Hooting occurs most in the early-evening and pre-dawn hours.

The male is the one that does the most territorial advertisement. Occasionally from two to five owls may be heard hooting in the evening, seeming to respond to one another. This is probably territorial hooting between males, since females are generally silent except for the few weeks of courtship. Most hooting humans do to attract or census owls probably elicits a territorial response from the male only.

Territorial conflicts are rare, with the birds generally staying clear of one another's territories. When conflicts do occur the birds may give a variety of loud vocalizations, such as squawks, barks, screams; they may also engage in fights, attacking each other with wings and talons. During aggressive interactions, owls may pose with plumage ruffled, wings partly spread, bills snapping, and swaying from side to side.

Following breeding, around August, when the young are independent, the adults wander outside territorial boundaries, possibly in search of better feeding grounds. Territories seem to dissolve at this time. In general, territories are again occupied in late fall.

Some birds do not occupy their territories until just before breeding in late winter or early spring. This is especially true of northern owls that may drift slightly south for the winter.

Great horned owls may have territories that overlap the ranges of hawks, but rarely do other large owls, such as the barred, live within their territory. This may be because of the danger of the great horned owl's killing or preying on the smaller owl.

Courtship

Main behavior: Hoots, Wing-spread, Bowing, mate-feeding
Duration: From winter to early spring

There is not a great deal known about great horned owl courtship and there are very few published accounts of what actually happens. Sometime after the male starts to define his territory in fall, the female joins him. In some cases the pair may remain together all year, and in other cases they may drift apart in the period after breeding and before territory formation.

Great horned owl courtship involves several behaviors, including: the approach of the male, calls, visual displays, allo-preening, and mate-feeding.

Courtship occurs most in the early evening, often just at twilight. The first stage is the male's hooting as he gradually approaches the female, landing on perches nearer and nearer to her. She may occasionally answer with her higher-pitched and shorter hoot. When within sight of her, he may do several types of visual displays and give an increasing variety of vocalizations. One display involves fluffing body feathers, partially spreading wings, and bowing repeatedly. He may also walk and hop about on the ground, or throw his head back and Bill-snap.

Eventually, he gets on the same perch with her. At this point she may behave aggressively toward him by Bill-snapping or fluffing out her feathers. If this is done, the male may renew his posturing and gradual approach.

If she is not aggressive, then he will sidle next to her and the two may engage in allo-preening, with each alternately pecking at the feathers around the other's bill and head region. This may reduce any aggressive tendencies on the part of both birds. During these close encounters a variety of vocalizations may be given by either bird. The pair may fly a short distance together, and both male and female may do Hopping and Bowing with a variety of calls.

Occasionally a male may bring prey to the female in a mate-feeding behavior. Then either she or both of them may eat it.

Mating may then occur, often taking place on the ground.

Once paired, male and female remain fairly closely associated, often roosting together during the day.

Nest-Building

Placement: Generally in trees, but also on cliffs
Size: Varies, since nest is not made by the owl
Materials: Usually sticks

Great horned owls do not build a nest of their own. Instead, they use an existing nest or other natural location. Where they nest depends on what is available in the area. When old nests of hawks or eagles are available they usurp these from the previous owners.

Competition with red-tailed hawks, whose nests they often use, is minor since the owls nest well before the hawks. Hawks finding their old nest occupied by a great horned owl rarely contest it, preferring to build a new one instead. If hawk nests are not present the owls may use the nests of crows, herons, or squirrels. In other cases the owls may use broken stubs of large trees or hollows rotted out of trees. In areas with cliffs they may nest on the ledges or within small caves. They have even been known to nest on the ground.

Generally, the only work the owl will do on the nest is to clear out a bit of old material; no new material is added except some feathers from the bird's breast. In cliff nesting the eggs are laid right on the ground.

Nests may be used from one year to the next, or the birds may switch to a different spot within their territory. The pair do not usually occupy the nest until a few days before the first egg is laid.

Breeding

Eggs: 1–4. White
Incubation: 28–30 days, mostly by female
Nestling phase: 6–8 weeks
Fledgling phase: 3 months
Broods: 1

Egg-Laying and Incubation

Soon after occupying the nest the female lays the first egg. She starts incubating after the first egg is laid, for it may be as early as February and the air temperature may be below freezing. The other eggs are usually laid at two-day intervals, although in some cases the interval is three to four days.

The number of eggs in a clutch varies from year to year. Since it has been observed that most of the owls in a given area vary their clutch size in the same direction in any given year, the birds may have some way of assessing available food and correspondingly adjusting clutch size. Most birds in the East have two eggs per clutch, while those in the Midwest and West tend to have three or four eggs. Nests in Florida have often been reported to have only one egg.

If eggs are destroyed, the birds may attempt a second clutch in either the same or a new nest. This usually occurs after about three weeks.

Incubation lasts for twenty-eight to thirty days, and although it is done primarily by the female, the male may also incubate as the female takes a turn hunting. The arrangement between male and female about how much each shares in incubation seems to vary with individual pairs. In most cases, the female incubates during the day while the male perches from a few to several hundred yards away. But there are records suggesting that the male may do daytime incubating as well.

In the early evening, one bird usually relieves the other at the nest; this is usually the male relieving the female. He may hoot as he approaches the nest. Upon landing at the nest, he may give a variety of vocalizations, including clucking sounds. One or both birds may do the Bowing display, and allo-preening may take place (see Courtship). Then the incubating bird flies off. Less elaborate exchanges at the nest may also occur, such as just a few hoots by the male.

The male or female may bring food to the nest during this phase and eat it alone or share it with the other bird.

Nestling Phase

Since incubation starts with the first egg laid, hatching is asynchronous and occurs over several days. Newly hatched owls are not much larger than baby chickens. They are covered with white down, cannot hold their heads up, and have their eyes closed. They lie on the floor of the nest and can peep only weakly, though sometimes loudly enough to be heard from the ground.

At two weeks their eyes are open, and you may see a few feathers unsheathed on their back or wings. At three weeks the wing feathers are about two inches long and the tail feathers one to two inches long. The birds still cannot move about very easily. During the fourth and fifth weeks the young gain mobility and can wander about the nest. After the third week they are no longer brooded by the parents.

During the sixth to eighth weeks the young leave the nest. Before taking their first short flight they perch on nearby branches. In a fair number of cases the young leave the nest early, possibly because of storms, disturbance, or a flimsy nest. In these cases the young usually perch on the ground at the base of a tree and the parents bring them food.

Prey items are brought to the young in the nest and their remains are often found rotting there if they are not completely eaten. The young cough up pellets and also back up to the rim of the nest and defecate over it. This may create whitewash on the vegetation below or on the nest rim.

Fledgling Phase

The fledgling phase lasts for another three months. During this time the young at first just stay perched in the territory and wait for food to be brought to them. When they are nine to ten weeks old they begin to attempt sustained flight and may follow after the parents and call loudly. They gradually develop their flying ability and learn how to hunt. At about five months they are ready to live on their own, and both parents and young disperse off the territory.

Plumage

DISTINGUISHING THE SEXES The male and female have similar plumage and can only be told apart by size or behavior. The female is larger than the male, but this may be hard to see unless the two are right next to each other. The difference in their voices is one of the best ways to distinguish between them. The male's voice is lower and more resonant and he does more of the hooting. The female's voice is higher, her calls are shorter, and she does less vocalizing.

DISTINGUISHING JUVENILES FROM ADULTS In the juvenile the throat patch is less white, the ear tufts are shorter, and the barrings are farther apart than in the adult. Juveniles may retain some downy plumage around the neck through the winter and into the next summer.

MOLTS Adults have only one annual molt, which occurs in mid-summer and is completed in mid-fall.

Seasonal Movement

Although great horned owls are generally nonmigratory, there are certain subspecies, especially those in the North, which to varying degrees move south during winter. Their movements do not seem to coincide with particularly harsh winters or the population cycles of their prey animals. Other great horned owls, even in the middle latitudes, may shift east or west for the winter.

There is also a period right after breeding when all great horned owls seem to leave their territories and move about until mid-fall, when the territories are reoccupied.

BobHines

Barred Owl
Strix varia

THE BARRED OWL IS AN EXTREMELY VOCAL BIRD AND IT IS NOT UNCOMMON to hear it hooting away even in the middle of a summer day. It is amazing how little is known about this fairly common owl. Even the length of its incubation period is still in question. In addition, there is scant knowledge of its visual displays, vocalizations, and territorial or courtship behavior.

A fairly common occurrence with barred owls is to have one show up in a city park during winter. These are probably males that have temporarily left their territories due to lack of food and have come into the city to feed on small birds and rodents. It is possible that the female, being the larger and more dominant bird, actually forces the male out of the territory, although this has never been witnessed.

It is our hope that, upon reading this summary of what *is* known about the barred owl, people will take an interest in looking more closely at this species and start to unravel some of the mysteries of its basic life.

BEHAVIOR CALENDAR

	TERRITORY	COURTSHIP	NEST-BUILDING	BREEDING	PLUMAGE	SEASONAL MOVEMENT	FLOCK BEHAVIOR
JANUARY	▓						
FEBRUARY	▓	▓					
MARCH	▓	▓	▓	▓			
APRIL	▓			▓			
MAY	▓			▓			
JUNE	▓			▓			
JULY	▓			▓	▓		
AUGUST	▓			▓	▓		
SEPTEMBER	▓				▓		
OCTOBER	▓				▓		
NOVEMBER	▓				▓		
DECEMBER	▓						

DISPLAY GUIDE

Visual Displays

Barred owl behavior has not been studied well enough to enable us to list specific visual displays. These owls seem to use forms of bowing and wing-spreading similar to those described for the great horned owl.

Auditory Displays

Hoot
Male or Female   *Sp Su F W*

A series of hoots given with a definite rhythm, often ending with a down-slurred note. Sounds like "hoo hoo hoohoo, hoo hoo hoohoo-awwww." Sometimes likened to the phrase "Who cooks for you, who cooks for you-all?"

CONTEXT: May function as a contact note between mates or as a territorial challenge. Also given during disturbances at the nest. May be heard in the daytime as well as at night. Most commonly heard during the courtship period and during late summer and fall.

Ascending-Hoots
Male or Female   *Sp Su F W*

A series of six to nine hoots, each slightly higher pitched than the previous one, ending with a down-slurred note as in the Hoot.

CONTEXT: Not known. Often given in circumstances similar to those of the Hoot.

Hoo-Aww-Call

Male or Female   *Sp Su F W*

A single, sharply descending call. May be repeated a few times at intervals of a minute or more.

CONTEXT: Not known. Possibly a contact note between members of a pair. Often given away from the nest site.

Ascending-Wail

Male or Female   *Sp Su F W*

A sharply ascending whistle, much like that made by humans when they blow through two fingers placed in the mouth.

CONTEXT: Not known.

Whether the above calls are related, gradations of the same call, or each a separately functioning call is not known.

Other calls by barred owls include a jumble of cackles, caws, and gurgles often given by two birds that are interacting. These may be given between mates or between neighboring territorial birds. Many of these defy any clear description and none of them have been studied well enough for us to understand their function.

BEHAVIOR DESCRIPTIONS

Territory

Type: Nesting, mating, feeding
Size: 1 square mile
Main behavior: Hoots
Duration of defense: Throughout year

Very little study has been done of barred owl territorial behavior. The owls generally respond quickly to playbacks (recorded Hoots or human imitations of their Hoot), often approaching the location of the playback, or approaching what is perhaps their territorial boundary if the playback is done outside their territory. Male and female both respond to playbacks and calls from strange owls and both seem to defend the territory. A variety of vocalizations are used during territorial encounters, but their significance is still unknown. Increased defense or advertisement of territories does not seem to occur at any particular time of the year.

The average territory size is about one square mile. Occasionally, only a smaller portion of this may be used during the breeding season. The territory may be expanded in winter due to lack of food; the owls are forced then to utilize marginal habitats for hunting. Males may also leave territories in winter and feed elsewhere, while the female remains on the territory. These males often show up in suburban areas or cities looking for food; they then return in the early spring to their mate and territory. Otherwise, both birds occupy the territory throughout the year.

Traditionally, the barred owl has been assumed to live primarily near wet areas, but new studies show no such preference exists. The birds do seem to prefer old woods that have a clear understory. Such habitats have trees old enough to have cavities in which the birds can nest and also provide good sight lines and access to prey on the forest floor. The adults generally have preferred feeding and roosting perches within the territory, and pellets of fur and bones of the prey they have eaten accumulate underneath.

Courtship

Main behavior: Hoots and possibly other behavior
Duration: February and March

Very little is known about the courtship behavior of our common owls, and among these, the least is known about that of the barred owl. The male and female tend to remain on the territory throughout the year, except in those cases when food on the territory is scarce and the male leaves for the winter months.

It is believed that the general pattern of barred owl courtship is similar to that of the great horned owl. The male approaches the female with a variety of vocalizations. Her voice, which is somewhat higher, is not heard as much in these encounters. Once the male is near the female he may do displays similar to those of the great horned owl, such as Wing-spreading and Bowing.

Courtship seems to occur in February or March. Once paired for the breeding season the two stay fairly close together, perhaps keeping in contact through various short calls and hoots as they move about their territory at night.

Nest-Building

Placement: Tree hole or cavity, usually located over 20 feet high; occasionally a nest of another bird is used
Size: Hole entrance must be 6 inches in diameter or greater
Materials: No materials are added to the nest, except possibly when other birds' nests are used; then moss or a few grasses may be added

Generally, the first choice of barred owls for a nest site is a cavity or hole in a tree. This cavity may be in the side of the tree where a limb fell off and the trunk then rotted out, or in the top of a dead stub where the inner wood is rotted out. Minimum cavity entrance size is about six inches in diameter for side-entrance holes and more like twelve inches for top-entrance holes. Average cavity depth in one study was about twenty inches. Cavities are usually lined with nothing but a few feathers from the owl.

When cavities are not available the birds generally use the old nest of a hawk, crow, or squirrel. The nests of red-shouldered hawks and Cooper's hawks are often used. Some slight alterations to the nest may be made by the owl, such as altering the rim a bit and lining the interior with a little grass, moss, or pine needles.

Occasionally, barred owls are reported to build their own nests if neither a cavity nor another bird nest is available. This is unusual among our owls and probably rarely occurs. These nests built by the owl are poorly made and may fall apart, leading to unsuccessful breeding.

The owls are generally constant to a nest site, using it for many years in a row. In one case a pair used the same cavity for nine consecutive years.

Nests are usually located twenty-four to eighty feet above ground. They are difficult to locate without following the owl to them, since there are rarely signs of pellets or whitewash around the nest tree.

Breeding

Eggs: 2 or 3. White
Incubation: Approximately 28–33 days, mostly by female
Nestling phase: 4–7 weeks or longer
Fledgling phase: Until late summer or fall
Broods: 1

Egg-Laying and Incubation

The eggs are usually laid every other day, occasionally at even longer intervals. Some incubation may begin with the laying of the first egg, especially at night, when one of the adults is always on the nest. During egg-laying, the parents may not remain constantly on the nest during the day. But once the final egg is laid, the nest is constantly attended except for brief moments that may occur in the morning and evening.

It is not known how much the male participates in incubation. If at all, its participation seems to be slight. As with the great horned owl, it may vary with different individuals.

There are very few records of the behavior of male and female barred owls during the incubation period. There is no record of prey remains found at the nest during this time, suggesting that male and female feed away from the nest. It is not known whether or to what extent the male may provide food for the female.

Even reports of the length of incubation are varied. The shortest time reported is about four weeks, but recent studies suggest it may be as long as five weeks. More study needs to be done.

Nestling Phase

The young hatch over a period of several days, showing that incubation starts before the clutch is fully laid. For the first two to three weeks the parents brood the young on the nest. They are brought food and it is ripped up and fed piece by piece to them by the parents. It is not known how the nest is kept clean of debris and droppings when the birds nest in tree holes.

In the first week, the young are covered with white down and their eyes are closed. In the second week, their eyes open and they may crawl about the nest. By the third week, new downy feathers replace their natal down and the beginnings of wing feathers can be seen. At any time from about four and a half to nine weeks after hatching the young may leave the nest. The large degree of variance may be explained by differences in how crowded the nesting cavity becomes. Earlier leaving may be a result of a small cavity's becoming too crowded.

In any case, when the young leave the nest, they are not able to fly. They crawl out of tree cavities by using their beak and talons to hold on to the bark. In fact, if they fall to the ground, they can climb trees that have rough bark, holding on with their bill and spreading their wings across the trunk as they take steps with their feet. In twenty minutes they can climb up nearly fifty feet.

The behavior of the adults during the breeding period varies with individual pairs. In general, barred owls are known to be fairly docile and will silently leave the nest when it is approached. In other cases, they have remained on the nest and not moved at all even when observers came very close. However, caution should be used around all birds of prey since they have powerful claws and can be aggressive, especially around their young. This is particularly true if the young are fledged and on the ground as you approach.

Fledgling Phase

When the young come out of the nest they are still flightless and generally just perch on nearby limbs. When the birds are about seven weeks old their main winter plumage starts to grow in. Between their twelfth and fifteenth weeks they start to fly and may leave the area of the nest with the parents. The young continue to be fed by the parents until late summer or early fall.

One of the more common calls of the young at this time is a hissing-squeaking sound about three seconds long that rises in pitch at the end. It may be repeated as often as two to three times per minute. The young can also make other chittering sounds.

After they develop flight, the young follow the parents about as the parents hunt. The young stay close to each other, usually perching in the same tree. They may also roost on the ground, usually among tall grasses, even after they can fly. By late summer, the young can catch prey on their own, but still remain with the parents and continue to call. They probably go off on their own in the fall.

Plumage

DISTINGUISHING THE SEXES The plumage of male and female barred owls is the same. The female is the larger of the two, but the difference in size may not be easy to see. One of the few other clues to the sex of the bird is the sound of its voice. The female's voice is slightly higher and more shrill than that of the male. The female also does the majority, if not all, of the incubation.

DISTINGUISHING JUVENILES FROM ADULTS Young and old birds look similar in winter except that the young bird is more reddish brown. The reddish tips of its feathers wear off during the winter, and by the juvenile's first spring, adult and young are indistinguishable.

MOLTS Barred owls have one complete molt per year starting in mid-July to early August and ending in late October to early November.

Seasonal Movement

Barred owls do not migrate. In winter, males may move off their breeding territories and spend the winter in other locations with better feeding resources. Occasionally, they even winter in suburban areas and cities where there are small birds or mammals on which to feed.

Eastern Screech Owl
Otus asio

WHEN FIRST HEARD, THE EERIE WHINNY-CALL OF THE SCREECH OWL CAN be a chilling experience, especially when one is out in the woods alone. But the little owl that makes the call is not threatening at all when seen. Our most common view of this owl is as it stares out from a tree hole or nest box in the middle of the day.

Although the screech owl is common in the country and suburbs, there is still a great deal to be learned about its behavior. Except for its Whinny-call and Monotone-call, its vocalizations have only rarely been recorded. Territorial interactions between screech owls can only be guessed about, and its courtship behavior has been recorded just a handful of times.

And yet, many people have gone out and continue to go out during the night with tapes of the bird's main calls to try to elicit a response from the owl, often for the Audubon Christmas Count. We need to move beyond censusing this popular owl and into understanding its life cycle and its interactions with its own and other species.

BEHAVIOR CALENDAR

	TERRITORY	COURTSHIP	NEST-BUILDING	BREEDING	PLUMAGE	SEASONAL MOVEMENT	FLOCK BEHAVIOR
JANUARY	▓						
FEBRUARY	▓	▓					
MARCH	▓	▓	▓				
APRIL	▓			▓			
MAY	▓			▓			
JUNE	▓			▓			
JULY	▓			▓	▓		
AUGUST	▓				▓		
SEPTEMBER	▓				▓		
OCTOBER	▓				▓		
NOVEMBER	▓				▓		
DECEMBER	▓						

DISPLAY GUIDE

Visual Displays

The visual displays of screech owls have not been sufficiently studied to enable us to decide which of their actions are discrete displays. However, brief accounts suggest that they may have bowing and wing-spread components to their courtship displays that are similar to the displays of great horned owls.

Auditory Displays

As with most owls, the male screech owl has a lower voice than the female, even though it is smaller in size. The male is also the one that most often gives the Whinny-call and Monotone-call, while the female is more likely to give a variety of hoots and barks, especially when defending the young.

Whinny-Call
Male or Female   *Sp Su F W*
A tremulous call that rises up and then descends.
CONTEXT: Most often given during territorial defense. Most commonly heard in late summer and fall. *See* Territory, Fledgling Phase

Monotone-Call
Male or Female   *Sp Su F W*
A tremulous call all given on one pitch—a monotone.
CONTEXT: Often given during interactions between male and female. Short versions may be given alternately between mates in a duet.

May function to advertise ownership of the nest hole by the male. Most often heard in late winter to early summer. *See* Territory, Courtship

Bill-Snap

Male or Female   *Sp Su F W*

A sharp snapping of the bill that makes an audible click.

CONTEXT: Usually directed at aggressors at close distance, such as at a possible predator at the nest or at songbirds mobbing a roosting owl.

BEHAVIOR DESCRIPTIONS

Territory

Type: Nesting, mating
Size: Area right around nest hole
Main behavior: Whinny-call, Monotone-call, chases
Duration of defense: Late fall to early spring

Most aggressive interactions between screech owls occur right around the nest site rather than over a larger area. Defense starts in late winter as the male claims one or more nest holes. He may announce his ownership through the Monotone-call.

The range of a screech owl depends on the concentration of food in the area. In rural or wild areas, a typical range may be from seventy-five to a hundred acres; in urban areas, where the owl's food (small birds, insects, rodents) is more prevalent, ranges may be as small as ten to fifteen acres.

Screech owls generally remain on their ranges throughout the year, only occasionally being forced to move further afield during winters when food is scarce. There is some evidence that screech owls are territorial on their range and defend it against neighboring screech owls. However, interactions are probably rare since each bird most likely becomes accustomed to its neighbor's presence and respects its range.

What are believed to be territorial interactions have been recorded several times. They may start with the birds' giving either the Whinny-call or Monotone-call. As the birds approach each other, a variety of other low aggressive sounds are made. After these initial calls, one owl may fly at the other and the two may fight, possibly falling to the ground. The calling and fighting may be repeated until one bird leaves. It is not known whether both male and female participate in territorial interactions.

Territorial defense may be most common in late summer and fall when young owls are dispersing and established adults are discouraging them from settling in their area.

Screech owls respond to imitated calls or recorded playbacks by approaching and sometimes answering. The responding bird is probably approaching what it thinks is an intruding owl. Screech owls may be nearby when playbacks are done yet not respond, in which case the playback may have originated just outside the territory of the bird. In general, only one member of a pair responds to the playback. It may be that this is usually the male and that males do the majority of territorial defense, as with the great horned owl. The owls respond most to playbacks during early fall, winter, and spring.

Courtship

Main behavior: Whinny-call, Monotone-call, mate-feeding
Duration: Late winter into the breeding phase

The beginning of a relationship between male and female screech owl centers on the nest hole, with the male claiming one or more nest sites in late winter and attracting the female to them. He then will start to bring all food to her, leaving it off in the nest hole or at a perch nearby.

As for actual courtship and mating displays, there are only a handful of recorded observations of these and they are most likely rarely seen. Much more study needs to be done. The scant data suggest that courtship displays start in February or March and involve the male's gradually approaching the female while giving the Whinny-call or Monotone-call.

As he gets nearer to her, he may do several visual displays such as Bowing or Wing-spreading. If the female is not aggressive toward him then he may move right next to her and the two may do allo-preening, nibbling at each other's feathers in the bill and head region. Just before copulation, the two may give the Monotone-call back and forth. They may sit right next to each other while continuing to call and, occasionally, allo-preening. They then may change their vocal behavior, with the male and female alternately giving shorter versions of the Monotone-call in a sort of duet — the female's call clearly higher than the male's. This calling is stopped when the male steps onto the back of the female and they mate. A second mating soon after the first may be preceded by fewer or no displays.

Nest-Building

Placement: In tree holes, bird boxes, or building cavities, from a few to over 30 feet above ground
Size: Variable
Materials: No materials are added to the nest

Screech owls use cavities for nesting, feeding, caching food, and roosting during the day. These may be natural tree cavities, such as those where inner wood has rotted out, or excavated holes made by woodpeckers; old flicker holes are often used. When tree holes are not available, screech owls will use bird boxes, such as those often supplied for wood ducks, or any crevice in outbuildings and barns. The owl adds no material to the nest, but just lays its eggs in the bottom of the hole. In spring and summer these cavities are most important for nesting. But throughout the year screech owls also use cavities for protection from predators and weather.

During the day, screech owls need to rest in a spot that is well hidden from hawks and mobbing birds. Where they roost during the day is most likely determined by the weather and the amount of available cover. In summer, when leaves are out, the birds can roost in all sorts of trees and not be seen. In winter, when it is cold and there are no leaves to hide them, they may prefer evergreens or tree holes and bird boxes for roosting and feeding.

Since the screech owl must be wary of larger owls at night, it often feeds within its cavity. Thus, the cavity may contain bits of food items and regurgitated pellets. The screech owl also frequently caches food items in its cavity, so a cavity may also contain uneaten birds or mice.

Breeding

Eggs: 4–6. White
Incubation: Average 27–30 days, by female only
Nestling phase: 4 weeks
Fledgling phase: 6–8 weeks
Broods: 1

Egg-Laying and Incubation

Egg-laying occurs in early to mid spring. The female stays in the nest hole for several days prior to laying her first egg. The first few eggs in the clutch are laid every one to two days and the remaining eggs every two to three days. Some incubation may start with the laying of the first egg. Only the female incubates the eggs and she is the only one to have a brood patch. The male may take a brief turn on the nest when she leaves, but he probably does not actually incubate.

During incubation the male brings all food to the female. Food items, often partially eaten by the male, are deposited in the nest hole or transferred to the female at a nearby perch. Food remains build up in the nest as do the female's regurgitated pellets and fecal waste. This makes the nest quite messy by normal standards, even before the young hatch.

Nestling Phase

The young hatch over several days and, as of this time, the male brings much more food. Food is dropped off in the nest and torn apart by the female, who then places bits in each nestling's mouth. Each evening, at about the same time, the female may leave the nest for a short period; she may also leave briefly in the morning.

In the first week the young owls have their eyes closed and are covered with fine, white, hairlike down. For the first two weeks, the female broods the young on the nest to help keep them warm during the period when they are developing feathers. As of the beginning of the third week, she may roost outside but near the nest box. The young are then developing feathers and their eyes are open.

Both parents now collect food for the young and may make as many as ten to seventy trips per night to the nest with food. High numbers of trips are generally required because small items, such as moths or other insects, are being brought. In some cases, food items may pile up uneaten, but they are later eaten by the rapidly growing young. After landing at a nearby perch, the adults swoop directly to the nest to transfer food and perch on the entrance with their head inside.

During the fourth week, the young leave the nest.

Fledgling Phase

During the last days of the nestling phase the parents feed the young less, and they may actually lose some weight. This creates a strong impetus to leave the nest. Once they are out the parents resume feeding them at a sustaining rate.

The young cannot fly effectively when they leave the nest cavity, but they can climb using their bill, talons, and wings. This enables them to move about the nest tree to some extent or even to climb back up if they fall down to the ground.

The parents roost near the young birds, and the female is very protective of them at this stage. She may even swoop down on you if you approach too closely. A week or so after leaving the nest the young may begin to follow the parents about as they hunt. Gradually the young hunt and feed on their own.

The family group may stay together until late summer or early fall. At this point the young disperse in all directions; most, however, do not move much more than a mile from where they were born. Some may move further; one record distance is a hundred miles. The adults remain on their range.

Plumage

DISTINGUISHING THE SEXES There are no differences in the plumage of male and female. In terms of behavior, it is the female who does all incubation and who has the higher-pitched voice.

DISTINGUISHING JUVENILES FROM ADULTS It is difficult to distinguish juveniles from adults.

MOLTS Screech owls have one complete molt per year, which starts in late July and is completed by mid-November.

POLYMORPHISM Eastern screech owls are generally polymorphic; two different color forms coexist irrespective of sex and age. These color phases are the red phase and the gray phase. Studies have shown that more gray-phase birds are found in the North and more red-phase birds are found in the South. In Florida, the birds are not classified as polymorphic, since many gradations between the two phases occur regularly. It is believed that selection may be occurring in the North, with red-phase birds being killed off by periods of extreme cold, to which they are not well adapted. This is still just a theory.

Seasonal Movement

The screech owl is not migratory. Adults stay on their breeding ground throughout the year, moving away only when there is a shortage of food. The young disperse in late summer and fall.

Ruby-Throated Hummingbird
Archilochus colubris

WHAT IS IT ABOUT HUMMINGBIRDS THAT HUMANS FIND SO FASCINATING? Is it their tiny size — a ruby-throated hummingbird weighs only a tenth of an ounce and measures three and three-quarters inches in length; their sparkling jewellike plumage — there are special structures in hummingbird feathers that create the iridescence; their amazing flight abilities — a ruby-throat can fly in any direction, including, briefly, backward; or is it their voracious appetite — a hummingbird can consume 50 percent of its weight in sugar each day (without getting fat!)? Whatever the reasons, hummingbirds are spellbinding to watch and a favorite of bird-lovers.

The easiest way to attract hummingbirds to your property is to plant a profusion of annuals, perennials, and shrubs that have nectar-rich flowers, especially red tubular flowers. After hummingbirds have started visiting your flowers, put hummingbird feeders near the flowers and the hummers will use them as well. Be patient; some hummers discover feeders right away and others take longer.

Watch the hummingbirds that come to your yard, for there is still much to be learned about their behavior. What are the patterns of interaction between male and female, adults of the same sex, and adults and young? Hummingbirds are very aggressive around food sources, and you will see much chasing and displaying, sometimes even directed at humans!

Ruby-throated hummingbird displays need to be better studied. Some writers feel that their displays function mainly to defend a food source and have little to do with courtship. Unlike most birds, which form a pair bond and raise the young together, male and female ruby-throats associate only briefly during mating, and then the female raises the young on her own. Exactly how male and female get together to mate has not been well documented.

In the United States, there are sixteen species of hummingbirds that breed. The ruby-throated hummingbird is the only species found in the eastern half of the country, although very rarely a western species is seen in the East. The ruby-throat breeds in southern Canada and throughout the East, the South, and the midwestern United States.

BEHAVIOR CALENDAR

	TERRITORY	COURTSHIP	NEST-BUILDING	BREEDING	PLUMAGE	SEASONAL MOVEMENT	FLOCK BEHAVIOR
JANUARY							
FEBRUARY					■		
MARCH					■		
APRIL	■	■				■	
MAY	■	■	■	■		■	
JUNE	■		■	■			
JULY	■			■			
AUGUST	■			■		■	
SEPTEMBER						■	
OCTOBER							
NOVEMBER							
DECEMBER							

DISPLAY GUIDE

Visual Displays

Pendulum-Arc-Flight
Male   *Sp Su*

Bird flies back and forth along the precise path of a wide arc, rising on each side from three to forty feet.

CALL: A loud buzz, supposedly made by the wings and tail, given at the lowest point in the arc; bird also may give grating squeaks

CONTEXT: Done by males during aggressive encounter with other males or females. It may have a courtship function. *See* Territory, Courtship

Vertical-Flight
Male or Female   *Sp Su*

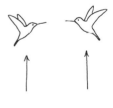

Two hummingbirds fly vertically up and down facing each other about one to two feet apart. Sometimes done in a seesaw manner with one bird at the top of its flight while the other is at the bottom of its flight. It is also reported that the birds may sometimes fly upward in a spiral while facing each other.

CALL: Twittering

CONTEXT: This may be an aggressive or, possibly, a courtship display. One report indicates that copulation took place on the ground after such a flight. *See* Territory, Courtship

Horizontal-Flight

Male or Female   *Sp Su*

Bird flies back and forth along a short horizontal path.

CALL: Twitter or hum

CONTEXT: We have seen this display given in aggressive encounters between hummingbirds and between hummingbirds and other species of birds. One writer reports that this functions as a courtship display and may be followed by a chase and then copulation. *See* Territory, Courtship

Tail-Spread

Male or Female   *Sp Su*

Bird fans its tail out.

CALL: Buzzing or twittering

CONTEXT: Done in aggressive and, possibly, courtship encounters. Done during Vertical-flights and Horizontal-flights.

Auditory Displays

Auditory displays have not been well studied or described in the scientific literature on ruby-throated hummingbirds. The birds have been reported to make humming or buzzing sounds, supposedly produced by the movement of their wings and tails. They also make a variety of twittering, grating, and squeaking sounds in many different situations.

BEHAVIOR DESCRIPTIONS
Territory

Type: Mating
Size: About ¼ acre
Main behavior: Pendulum-arc-flight, chases
Duration of defense: Summer

Hummingbirds behave aggressively toward one another much of the time. Under what conditions they actually become territorial and defend an area, and for what reasons, needs more study. There are times when hummingbirds may feed together at an abundant food source such as a large flowering tree and be somewhat tolerant of one another. At other times, hummingbirds may remain in, and defend, a specific area that includes nectar-rich flowers.

Male hummingbirds usually return from migration ahead of the females and seek out areas with nectar-rich flowers, such as gardens, where they may establish a territory of a quarter acre or more. When the flowers have finished blooming, the male may leave and become territorial around a new patch of flowers. Territorial behavior functions mainly to defend a food supply from other male and female hummers, although mating with females does take place within this area.

One male that was studied was seen to frequent a quarter-acre flower garden, from which he would chase other male and female hummers, often pursuing them beyond the boundaries of his territory. Bees and sphinx moths were also chased. He had favorite perches from which he would watch for intruders. His behavior consisted of flying directly at other hummers and chasing them, and, if they did not immediately leave, he would also display with Pendulum-arc-flight.

Female hummingbirds are territorial in the immediate vicinity of their nests, which are usually not near the male's territory. Under certain conditions, females may also be territorial around a food source. We have had female hummingbirds set up territories in our yard from July to September, coinciding with the blooming of our large patch of cardinal flower *(Lobelia cardinalis)*.

Each year a dominant female has kept vigil over the garden from favorite perches. She has flown at and chased both other female ruby-throats and immature males and females, although we have not seen her chase an adult male. She has also chased bees and butterflies and gone after other species of birds such as titmice, chickadees, goldfinches, even blue jays. We have seen females direct the Horizontal-flight display at other females, immature hummers, and other species of birds.

It was interesting to watch the reactions of other hummingbirds that tried to enter the garden. Sometimes they were not immediately noticed by the dominant female, but once spotted, they were persistently chased, often well beyond the boundaries of our property. Some intruders were persistent and were chased for hours. On several occasions the hummingbirds actually fought and fell grappling to the ground. Other intruders left promptly. Sometimes other hummingbirds would feed on flowers in the front of the house, which the dominant female could not see from her perch. Other hummingbirds that visited our garden would often enter and leave by the same route each time.

Courtship

Main behavior: Pendulum-arc-flight, Horizontal-flight
Duration: Before egg-laying

Hummingbirds do not pair for any length of time. They are promiscuous and only have contact briefly during mating. Then the female goes off by herself to raise the young. The male will mate with other females. It is not clear which displays of ruby-throats have a courtship function, or whether most of their behavior is aggressive and the male just overpowers the female during mating.

There have been observations of males doing Pendulum-arc-flight in response to females, following the path of an arc and making a loud buzzing or humming sound as they pass over her head. Other observations have males and females doing Vertical-flights together, with mating taking place on the ground after one of these flights. One writer listed Horizontal-flight as preceding copulation. There are only a few reports of mating being observed, and some writers think that what has been reported as mating was just aggressive behavior.

Nest-Building

Placement: On a small limb, frequently one that is lichen covered. Nest height ranges from 4 – 50 feet but is usually 10 to 20 feet off the ground in deciduous or coniferous trees
Size: Outside diameter, 1 – 1¾ inches; height, 1 – 2 inches; inside diameter, ¾ – 1 inch
Materials: Bud scales, lichens, and plant down held together with spider silk

The nests of ruby-throated hummingbirds are tiny, fascinating structures, no bigger than a half-dollar, and with their lichen camouflage may look like a knot on a tree limb. The female alone builds the nest, which may take from one to ten days. It is often on a downward-sloping limb, protected by overhead foliage, and the nest is frequently near or over water, such as a brook.

First a foundation of bud scales — the winter covering of flowers and leaves — is attached to the limb with spider silk. Then lichens are added to the outside, followed by a lining of plant down. The female tucks material in on the inside and works her body around and up and down, shaping the nest. She may continue adding lining during egg-laying and incubation.

The nest is capable of keeping out moisture. Since it is made of flexible material, it may stretch as the young grow.

Breeding

Eggs: Usually 2. Pure white
Incubation: About 16 days, by female only
Nestling phase: 14–31 days
Fledgling phase: Up to 34 days
Broods: 1 or 2

Egg-Laying and Incubation

The tiny white eggs are laid in the morning and there is an interval of one day between layings. Egg-laying may start before the nest is completed. The female usually starts incubating after the last egg is laid. She incubates from 60 to 80 percent of the day, depending on the weather. If it is cold she will sit tightly, but in very hot weather she may just stand over the eggs and shade them. When she wants to change position on the nest, she lifts herself straight up and, while hovering, turns in the air, then lowers herself.

Ruby-throats may sometimes have two broods. If there is a nesting failure, they will renest. Sometimes old nests are reused but usually a new nest is built.

Nestling Phase

The newly hatched young are no bigger than peas and dark gray with just a few downy feathers. They have short yellow bills. The female must brood them until they are able to maintain their own body temperature at about twelve days of age. The young do not grow down, but develop pin feathers after only a few days. This is the plumage they will wear until after fledging.

The young are fed from one to three times every hour. The mother alights on the nest and puts her bill straight down into their gaping mouths, then in a series of pumping motions she feeds them nectar and insects that are regurgitated from her crop. As the young grow and their bills get longer she may thrust her bill into theirs at a right angle.

Feces of the young are removed by the mother or eaten. (One observer reported that the mother at one particular nest placed the droppings of the young in a line on the branch just above the nest.) When the young are large enough, they back up to the edge of the nest and expel their feces over the side.

The young in the nest preen their feathers with their bills and feet. They may extend their tongues and touch nearby vegetation in an exploratory way.

Where there is abundant food and the weather is milder, the young may grow faster and leave the nest earlier. That perhaps is why there is such a range of lengths recorded for the nestling phase.

Fledgling Phase

Several days prior to leaving the nest, the young will lift up their wings and begin exercising them. Just prior to leaving, they will anchor themselves to the edge of the nest and rapidly vibrate their wings. Fledging occurs in the morning. They just lift themselves off the nest and hover in the air, and may be able to fly fifty feet or more. At first they may have a little trouble landing.

They usually stay in the vicinity of the nest for several days. There is one report of fledglings making a particular call, a shrill and far-reaching whisper. The female may continue to feed the young for some time, although the young explore many flowers and come to learn which ones are rich in nectar. The female may also bring the young to hummingbird feeders.

There is a record of one female ruby-throated hummingbird tending two nests at the same time. The nests were on the same branch, three and a half feet apart. The female alternated feeding one nearly full-sized young in the first nest with incubating two eggs in the second nest. She also chased robins, house wrens, and catbirds from the nest tree. All three young successfully fledged. The two second-nest young were last recorded feeding in a nearby flower garden and perching in a tree near the nest site nine days after fledging.

Plumage

DISTINGUISHING THE SEXES The adult male has a brilliant ruby red throat. In poor light the throat may appear black. His forked tail is blackish on the sides. The adult female has a white, rather than red, throat. The female also has a more rounded tail, with white on the corners.

DISTINGUISHING JUVENILES FROM ADULTS The young male and female resemble the adult female; in particular, they have white throats and white tips on the outer tail feathers. In August and September some young males begin to acquire a few ruby feathers on their throats.

MOLTS Ruby-throated hummingbirds have one complete molt per year, which may take place during fall migration and continue through February and March on their wintering grounds. The young males then acquire their full red throats.

Seasonal Movement

Hummingbirds begin their migration in fall. Sometimes numbers of them stop temporarily at a good food source, where they may be aggressive to one another. They migrate during the day and usually fly singly. We have stood atop a two-thousand-foot peak in Massachusetts in September and, while watching migrating hawks, have seen hummingbirds pass right over the top of the mountain, one every several hours.

Ruby-throats winter in Mexico and Central America. They fly nonstop across the Gulf of Mexico, able to store enough fat beforehand to make this spectacular over-water crossing.

Spring migration occurs in April and May, the birds moving north as the flowers they feed upon come into bloom.

Pileated Woodpecker
Dryocopus pileatus

THE PILEATED WOODPECKER IS OUR LARGEST WOODPECKER AND ALWAYS A thrill to see. In the South, it is a common bird often seen in suburban and rural areas. In the Northeast, it became very rare at the turn of the century due to the almost total clearing of the forests to create farmland.

In the last hundred years, as farmland in the North has been abandoned and allowed to regrow into forest, the pileated woodpecker has gradually returned to areas where it was once plentiful. Since the pileated feeds primarily on carpenter ants that live in the central portions of large trees, it takes many years for a growing forest to come to have trees old enough to support a pair of these birds. The birds are also much more wary of humans in the North and thus not as easy to watch.

Much of pileated woodpecker behavior is similar to that of other woodpeckers, such as the downy and hairy woodpeckers, and the common flicker, whose lives are detailed in volumes one and two of this series. Many of their displays are almost the same, such as the Bill-wave, Crest-raise, Threat-display, Drumming, and the Awoick-awoick-call. In some ways, the pileated is just a much larger bird doing all of the same behaviors.

Yet the Drumming of the pileated, which can be heard from a great distance, can easily be distinguished from that of other woodpeckers. It is generally lower pitched, but also is distinctive in that it gets softer at the end. Hearing this Drumming is often the best way to locate pileateds, since it is often done and the sound carries so well.

The pileated woodpecker ranges over a very large area, so that as a behavior-watcher you will have to depend on picking up bits and pieces of the bird's behavior. Through reading the following summary, you can come to understand how these pieces fit into the bird's year-round behavior.

BEHAVIOR CALENDAR

	TERRITORY	COURTSHIP	NEST-BUILDING	BREEDING	PLUMAGE	SEASONAL MOVEMENT	FLOCK BEHAVIOR
JANUARY	▓						
FEBRUARY	▓						
MARCH	▓	▓					
APRIL	▓	▓	▓	▓			
MAY	▓	▓		▓			
JUNE	▓			▓			
JULY	▓			▓			
AUGUST	▓			▓	▓		
SEPTEMBER	▓				▓		
OCTOBER	▓				▓		
NOVEMBER	▓						
DECEMBER	▓						

DISPLAY GUIDE

Visual Displays

Bill-Wave

Male or Female   W Sp Su F

Head is raised, sometimes held over back, and swayed from side to side in a forty-five-degree arc. Bill may be slightly open. Tail may be swung to same side as bill. Occurs while bird is perched on trees or on ground.

CALL: Awoik-awoik-call

CONTEXT: Occurs between birds of the same sex in territorial conflicts, and between male and female in courtship or close encounters between the pair. *See* Territory, Courtship

Crest-Raise

Male or Female   Sp Su F W

Red crest is raised to varying degrees.

CALL: Kek-kek-call, G-waick-call, or none

CONTEXT: Occurs during moments of excitement or alarm, such as during defense of the nest hole from squirrels. *See* Territory, Breeding

Threat-Display

Male or Female   Sp Su F W

Bird extends its wings to the side, showing the white undersides of the wings. Another variation of this may be a repeated flicking of wings. Bird may peck at other bird or nearby objects.

CALL: Kek-kek-call or none

CONTEXT: Usually done toward an intruder, such as another pileated woodpecker or other species of bird. *See* Territory

Auditory Displays

Cuk-Cuk-Call

Male or Female   *Sp Su F W*

A series of ten to fifteen low-pitched short notes, not as loud, fast, or high-pitched as the Kek-kek-call, but still carrying long distances. The most common call of the pileated woodpecker.

CONTEXT: Usually occurs during interactions between a mated pair. May be used as a contact call between male and female as they move about the territory. Often given when the birds are feeding separately. Given in response to hawks, including the red-shouldered hawk, Cooper's hawk, and sharp-shinned hawk. Sometimes denotes excitement. *See* Courtship, Territory

Kek-Kek-Call

Male or Female   *Sp Su F W*

A series of six to eight high-pitched, loud notes with the last note in the series lower pitched. Often followed by bursts of Drumming. A loud and penetrating call that carries long distances.

CONTEXT: Given in long-distance communication between a pair, also during moments of alarm. The other member of the pair may answer it with the same call or Drumming. Heard most in winter and spring. *See* Courtship

Awoick-Awoick-Call

Male or Female   *Sp Su*

Sounds like its written description. Easily distinguished from other calls by its context.

CONTEXT: Given between two birds when they are interacting at close range. Often accompanies the visual display of head-waving. Can be exchanged between members of a pair during courtship, or between territorial competitors. Most common in breeding season. *See* Territory, Courtship

Drumming
Male or Female   *Sp Su F W*
Rapid volleys of pecking on a resonant trunk or limb. Lasts two to three seconds and gets softer near the end. Sometimes repeated every forty to sixty seconds, four to seven times in a row. Sometimes done at a faster rate.
CONTEXT: Males drum most. Drumming can be heard in any month but is heard least in fall. Done at a faster rate by unmated birds looking for a mate. *See* Courtship, Territory

Tapping
Male or Female   *Sp Su*
Soft taps given from inside or just outside the nest hole. Only heard if you are near the nest.
CONTEXT: Given as birds change places at nest hole during excavation and incubation, and in general, when one bird has just arrived at the nest. *See* Nest-Building, Breeding

G-Waick-Call
Male or Female   *Sp Su*
A loud, shrill scream, given one at a time.
CONTEXT: Given in moments of alarm near the nest and during conflicts with other birds near the nest. *See* Territory

BEHAVIOR DESCRIPTIONS

Territory

Type: Nesting, mating, feeding
Size: 150–200 acres
Main behavior: Drumming, Bill-wave, Threat-display
Duration of defense: Throughout year

Pileated woodpeckers roam over large areas and may defend a territory as large as one hundred fifty to two hundred acres. This size varies greatly depending on the availability of food and proper nest-hole and roost-hole trees. Conflicts in fall and winter may occur near roost holes for which the birds compete. Conflicts in spring may involve an intruding bird challenging a pair for access to a mate. And conflicts in summer may occur at or near the nest hole where the birds are raising young.

Aggressive interactions between pileateds involve the use of several displays as well as direct pecking at each other. Visual displays include Bill-waving and the Threat-display; auditory displays include the Awoick-awoick-call and G-waick-call, or the interaction may be silent. Occasionally, Kek-kek-calls and Drumming are heard during conflicts, possibly as one member of a pair calls its mate to the site of the conflict.

During a typical encounter both birds may hitch around tree trunks, strike at each other, and do these displays. This may occur on trees or on the ground, and the birds can be quite oblivious to observers during the interaction. Most conflicts are between birds of the same sex, except during the breeding period when either male or female will evict a stranger from the area of the nest.

Conflicts with other species include those with squirrels, other hole-nesting birds, birds of prey, and tree-climbing snakes that may rob the nest. In general, squirrels are dominant over pileateds; even so, when squirrels are near their nest trees pileateds may try to attack them while doing the Threat-display or Crest-raise. The squirrel may fight back. If the squirrel decides to take over the nest it usually can.

Certain species of snakes can climb trees and eat the eggs or young of hole-nesting birds, including those of the pileated woodpecker. If the snake is large, the pileated may not be able to discourage it from going into the hole. Again, you may see the pileated use the Crest-raise and Threat-display during these encounters.

Other smaller hole-nesting birds, such as other woodpeckers, bluebirds, or wrens, may inspect a pileated nest, but they are soon routed when the larger owner arrives.

Habitats of pileated woodpeckers are often sparsely populated; the birds then have such large ranges that conflicts with other pileateds may be only rarely seen. How much conflict occurs depends on the density of birds and the amount of suitable habitat available.

Courtship

Main behavior: Drumming, Kek-kek-call, Bill-wave
Duration: Spring

Lone males in late winter or early spring may advertise for a female by Drumming frequently, twice per minute, from many different drumming posts around their range. Going from one post to another, the male may fly above the treetops and may give the Kek-kek-call. All of this makes him very conspicuous.

Mated pairs remain together on their range throughout the year. In spring they usually go through a courtship phase during which there is increased Drumming between the two, and possibly some close interactions involving Bill-waving. In these cases Drumming occurs less often than when a lone bird is seeking a mate—only about once per minute.

Copulation occurs before and during the period of egg-laying. During copulation the female perches across the branch, as do many other woodpeckers, and she may give Awoick-awoick-calls as the male approaches. Copulation usually occurs near the nest hole.

As with other woodpeckers, there are sometimes triangular encounters among pileateds, when an intruding bird challenges a mated pair and tries to steal a mate for itself. In these cases, the conflict is between the intruder and the member of the pair that is of its own sex, that is, male versus male or female versus female. The other member of the pair may just remain in the area, instead of flying away, and does not join in the conflict. G-waick-calls and Awoick-awoick-calls may be given and Bill-waving and Threat-displays may be used during these encounters. Chasing and pecking at each other are also common elements in these conflicts. Usually the intruder is routed.

Pileated woodpeckers spend the night in roost holes, each bird in its own hole. Sometimes these are old nest holes from previous breeding seasons; other times they are separate roost holes that may be used many years in succession. Roosting holes and nesting holes are usually in the same general area of the birds' territory.

Throughout the year a mated pair get together each morning after leaving their respective roost holes. This is usually just around sunrise. The first one to emerge usually gives Kek-kek-calls or Drumming, and this is soon answered with the same sounds given by the other bird. One then flies over to the other and they begin their day of foraging together. Pileateds often stay within forty to fifty yards of each other throughout the day, keeping in contact mostly with Cuk-cuk-calls given as they fly to a new location or just after they land. Occasional Drumming and Kek-kek-calls may also be given.

During the hour before sunset both birds return to their roost holes. Although they may return silently, sometimes they give several Cuk-cuk-calls as they fly to roost.

Nest-Building

Placement: Excavated hole in dead wood, 15–70 feet off the ground
Size: Hole is about 3½ inches in diameter; depth of cavity 10–24 inches
Materials: No materials added, although grains of sand and pebbles have occasionally been found in nests

Pileated woodpeckers generally make their nest in dead wood. The hole is anywhere from fifteen to seventy feet above ground. A tree has to be fairly large to accommodate a pileated nest, and thus in some young forests there may be only a few suitable trees. More than one hole may be started before a final nest is completed.

Finding a tree that is large enough and rotted enough can be hard. The birds may put a lot of energy into a hole and then find the inner wood too hard. Because of this, there may be competition between neighboring pairs around potential nest sites. Two or all four birds may interact as in territorial encounters.

Both male and female excavate, but the male generally does substantially more excavation than the female. A bird that is excavating may Drum or give the Kek-kek-call from the nest tree to attract its mate. When the two exchange places for excavating, the bird leaving may Tap at the nest hole entrance. Tapping may also indicate attachment to a nest site.

Complete excavation may take only a few days. The nest can usually be located by looking for the profusion of large wood chips below the tree. Many chips from inside the nest are also carried fifteen to twenty feet away from the tree and dropped.

Breeding

Eggs: 3 – 5. White
Incubation: 15 – 16 days, by male and female
Nestling phase: 3 – 4 weeks
Fledgling phase: Up to several months
Broods: 1

Egg-Laying and Incubation

As far as is known, the female lays one egg per day until the clutch, which usually contains four, is complete. Incubation is believed to start after the laying of the last egg. Copulation may be seen during this period of egg-laying and slightly before.

The parents are extremely attentive to the nest; they rarely leave it unguarded for more than a few minutes. In general, the incubating bird does not leave until the other shows up right at the nest hole. The bird who is not incubating remains fairly near, feeding in the woods, occasionally giving Cuk-cuk-calls.

During changeovers the approaching bird gives Cuk-cuk-calls. The incubating bird pokes its head out to see its mate. The other bird then flies to the nest tree. Often the incubating bird Taps from within the nest and then leaves. The other bird enters, may peer out of the hole briefly, and then settles in to incubate.

Incubation periods last one to two hours. During the day the male often incubates more than the female. The male also is always the one to remain on the nest at night. Thus, altogether, the male does far more incubation than the female.

Nestling Phase

It is hard to tell when the young hatch, because the parents brood the young all of the time for the first five days and much of the next five to nine days. Since they feed the young by regurgitation and do not remove fecal sacs during the first few days, there are no external signs of the hatching.

When the parent arrives at the nest to feed the young, it perches at the entrance, then leans down and in while still holding on to the rim of the nest. Its rear ends up against the top of the nest entrance; its head is straight down feeding the nestlings. After feeding, that parent may brood the young until the other parent arrives with food. Feedings occur irregularly but can be as far apart as one hour in the early stages and as far apart as two hours in later stages.

When the young are fifteen days old they can climb to the nest entrance and stick their heads out. The nestlings are quiet until a parent approaches, then they give a "chrr-chrr" call. The parent puts its bill into the throat of the young bird and pokes vigorously as it regurgitates food.

For the first few days after hatching the parents probably eat the fecal sacs of the young, but after that they can be seen carrying them away from the nest to be dropped elsewhere; several fecal sacs may be carried at a time.

A renewal of some courtship activities may be seen and heard during the later stages of the nestling phase. The parents may Drum or give the Awoick-awoick-call, and may Bill-wave upon coming close together.

The nestling phase lasts three to four weeks.

Fledgling Phase

Even though the young have been cooped up in the nest hole for several weeks and have had no chance to exercise their wings, their first flight from the nest can be almost a hundred yards. There may be more Drumming by the parents in the vicinity of the nest just before the time of fledging. Its function is not known.

Once the young leave the nest they use a short version of the Kek-kek-call as a contact note and a way for the parents to locate them. The young may also fly after the parents to get food. The young may still be fed in part by the parents for three months or more.

Plumage

DISTINGUISHING THE SEXES The male has red plumage extending from his crest to the base of his upper bill and he has red on his "mustache" line next to his bill. The female has red plumage on her crest only and not on her forehead or "mustache."

DISTINGUISHING JUVENILES FROM ADULTS Juveniles are slightly grayer than the primarily black adults and may have more streaking on their throats. Sexes distinguished as in adults.

MOLTS Adults undergo a complete molt of all feathers starting in late summer and continuing into fall.

Seasonal Movement

In general, pileateds are not migratory; pairs remain on their breeding territories throughout the year. However, there are some reports of increased populations of pileateds in winter in certain areas and of some showing up in new places in fall. These may be young dispersing from their area of birth and looking for their own areas in which to settle.

Purple Martin
Progne subis

SAY "PURPLE MARTIN" AND MOST PEOPLE'S FIRST THOUGHT IS OF MARTIN houses, the many-compartmented bird houses set on tall poles. This is because attracting purple martins through nest boxes has become such a popular pastime. Obviously, the birds have not always nested in human-made boxes.

Originally, purple martins nested in just about any cavity they could find — rock crevices, natural tree holes, and abandoned woodpecker holes. In western states this is still the case; in the Southwest the birds live in holes in giant saguaro cactuses. Although they may occasionally live as single pairs in the West, they are more often found in small colonies of about three to twenty pairs.

In the Midwest and East the birds nest almost exclusively in human-made nest boxes. This is a habit that may have been started by the American Indians, who hung gourds up on poles to attract the birds. One possible reason for this may have been that they wanted to protect their drying hides and game from hawks, vultures, and other large birds, since the purple martins are aggressive nest-site defenders and will readily chase off these larger birds. Besides nesting in human-made boxes, martins today may also use crevices in buildings, bridges, or other human structures.

A good question is, Do purple martins really like to live as close together as they do in the nest boxes? As you begin to watch their behavior you can gauge for yourself. It is clear that males and females are very aggressive at first toward their neighbors and that males will defend more than one nest hole even though they may not use the extra ones.

In any case, the nest boxes provide us with excellent opportunities for closely watching these birds' marvelous behavior and getting a clear look at their entire breeding cycle. If you have martins near you, don't miss the chance to enjoy their behavior.

BEHAVIOR CALENDAR

	TERRITORY	COURTSHIP	NEST-BUILDING	BREEDING	PLUMAGE	SEASONAL MOVEMENT	FLOCK BEHAVIOR
JANUARY					▓	▓	▓
FEBRUARY						▓	▓
MARCH						▓	▓
APRIL	▓					▓	▓
MAY	▓	▓	▓				▓
JUNE	▓	▓	▓	▓			▓
JULY	▓			▓	▓		▓
AUGUST	▓			▓	▓		▓
SEPTEMBER						▓	▓
OCTOBER						▓	▓
NOVEMBER					▓		▓
DECEMBER					▓		▓

DISPLAY GUIDE

Visual Displays

Horizontal-Threat

Male or Female   *Sp Su F W*

Bird holds its body horizontal; bill may be open and crest raised, and wings and tail may be flicked repeatedly. Bill may also be audibly snapped.

CALL: Song or None

CONTEXT: Given in territorial defense around the nest hole, but also in other situations of extreme aggression.

Nest-Hole-Flight-Display

Male   *Sp*

Male flies away from the nest hole in a large circle and on the return flight starts a steep dive toward the nest hole with wings flapping below the body line. Upon entering the hole, he immediately turns around and, with head protruding out of the hole, sings.

CALL: Song

CONTEXT: Done by male that is claiming a hole and trying to attract a mate or other birds to the nest area. *See* Territory

Flight-Display

Male   *Sp Su*

Bird flies in a labored way with back humped up, head down, and tail compressed into a single spine and held down. Position may be maintained after perching with wings drooped.

CALL: None

CONTEXT: Not fully understood. Has been suggested that it may be either a submissive display given by males defeated in an aggressive encounter or a sexual display.

Auditory Displays

As many as ten different vocalizations of purple martins have been recorded. Below are the most common ones that you will be able to hear and distinguish in the field.

Song
Male or Female   *Sp Su*
Several pairs of notes followed by a guttural warble or grating sound. The only long, complex vocalization of the purple martin. Lasts from two to six seconds and may be given flying or perched. Female version usually shorter and may not include grating sound.
CONTEXT: Used during courtship, copulation, forced copulations, and any other male/female interaction. Given when one bird replaces the other at the nest and when the pair rejoin after being separated. May be given along with the Horizontal-threat display in aggressive encounters. *See* Courtship

Cher-Call
Male or Female   *Sp Su F W*
Most common call of martins on the breeding grounds. A short harsh sound much like the written "cher." Given singly as in "cher" or doubly as in "cher cher."
CONTEXT: Given in all kinds of situations. Usually accompanied by flip of the wings or

body. Given as birds approach the colony, as they perch at the nest site, and during mild alarm. May be individually distinct, helping members recognize each other. Also given before dawn while birds are still in the box.

Zweet-Call

Male or Female   *Sp Su F W*

A short, one-syllable call with a definite "eee" sound. Sometimes descending slightly as in "zeert." Usually given in flight.

CONTEXT: General alarm call that may cause the other birds in the colony to fly up and circle. Given toward any intruder, including predators or other species competing for nest holes.

Hee-Hee-Call

Male or Female   *Sp Su*

A series of four to ten hoarse, high-pitched sounds much like the written sound "hee hee hee." Given at a rate of about four per second.

CONTEXT: Not that common. Given mostly during territorial fights between males. Occasionally given by males during Nest-hole-flight as they attract females to their territory. *See* Territory, Courtship

Juvenile Call

A single-syllable call, harsh and short, given about two or three times per second. Only call of birds in late-nestling and early-fledgling stages. Given when the parent arrives with food, when the young bird is in danger, or whenever the young move from one location to another. *See* Fledgling Phase

BEHAVIOR DESCRIPTIONS

Territory

Type: Nesting, mating
Size: Immediate area of nest hole(s)
Main behavior: Chases, song
Duration: From arrival on breeding ground until well after breeding

Male martins are generally the first to arrive at the breeding site. On their first day they may spend only a short time at the site, going in and out of holes, visiting nearby nesting sites, and, possibly, giving bits of Song. Then they may leave to feed for the remainder of the day.

How continuously the birds stay at the nest site upon first arrival depends on the severity of the weather. A cold spell may make them leave the area for a day or so until it warms up.

After repeatedly going in and out of many different holes, males gradually choose one or more nest holes and defend them against other male martins.

A territorial male often sings from the ledge in front of his nest holes or from a nearby perch, such as the top of the box (although the top of the box is never defended). Song increases in intensity as other martins fly nearby. The male may also do the Nest-hole-flight, in which it makes a long circular flight, returning to the hole with a steep dive and flutter flight.

When an intruding male challenges a territorial male, he may at first be met with the Horizontal-threat display. Then the intruder may be chased and the two may fight by facing each other in midair and locking claws. Sometimes the birds fall to the ground and continue fighting. The Hee-hee-call may be heard during these fights.

The territorial male may also retreat to one of his holes and do the Horizontal-threat display or peck at the intruder as it approaches.

Once territories are established, neighboring males no longer fight with each other and rarely go onto each other's territory.

Neighboring territory holders may even occasionally drive away intruders from each other's territories when the neighbor is away.

At the start of the nesting season male martins generally defend more nest holes than they eventually use. In one study, the number of nest holes defended early in the season averaged from eight to twelve. These were adjacent holes or holes on adjacent floors on one side of a martin nesting box. Males arriving later in the season started off by defending an average of three or four nest holes. As the season progresses, fewer holes are defended; by the end of the breeding season the number of holes defended averages about two per male.

It has been suggested that the defense of extra nest holes enables the male to be polygamous if the opportunity arises. In martins, polygamy seems to be attempted by about 20 percent of males and successfully undertaken by about 5 percent.

Martins generally sleep in one of their nest holes and defend their other holes at dusk before retiring. Intruding males without territories also look for nest holes to sleep in and usually arrive right at dusk, trying to enter unoccupied holes quickly and without being noticed by territory holders who might chase them away.

Males will let females enter their territory. Once females arrive and pair with a male, they also defend the territory from other males and females. Once the female starts to incubate she generally defends only the nest hole in which she is actually nesting.

Nest-hole defense continues throughout the breeding period.

Courtship

Main behavior: Song, chases, entering nest holes
Duration: From arrival of female to the beginning of incubation

Once males have claimed one or more nest holes, females arrive and begin to inspect them. A male may encourage a female to inspect his nest holes in at least three ways. He may sing louder and more frequently; he may rapidly go in and out of one or more of his nest holes; or he may do a Nest-hole-flight-display. Upon entering his nest hole after this flight, he immediately turns around and sings loudly from the hole entrance. Most courtship behavior occurs in either the early morning or early evening.

A female seems to inspect all available nest holes, whether they are guarded by a male or not. It seems that her choice of nest hole determines which male she will pair with. Once paired the two birds will do most activities together, such as preening, feeding, and flying, and will greet each other with Song when joining after being apart.

When a female looks into a nest already claimed by another female, the two may fight. The new female may then try to claim another hole in the same male's territory. This results in polygamy, with the male forming a bond with both females, protecting them from other males, and guarding both their nests when they leave on feeding trips during the incubation period. Although polygamy may be attempted by about 20 percent of males, in only 5 percent does the pairing last through the fledgling phase for both females. In the other 15 percent, the young may not complete development or the female may get the help of a different male.

When females first arrive at the breeding ground they usually sleep in a hole belonging to the male with which they will eventually mate. When a male has claimed only one nest hole, he and his mate will share it for the night. When a male has claimed more than one nest hole he may enter and leave several before settling down. He and his mate may sleep in the same or separate nest holes; generally, the longer the pair have been together, the greater the chance they will share the same nest hole for sleeping.

During nest-building the pair tend to sleep together in the hole in which the nest is being built. Copulation probably takes place in the nest hole at night before and during the egg-laying period. This is because copulation outside the nest box causes other males to approach, interfere, and make force-mate attempts. The birds vocalize in the latter part of the night from within the box, and continue until dawn. Whether this has anything to do with courtship or copulation is not known.

During the breeding season you may see two types of sexual

chases: pair-chases and rape-chases. Pair-chase can be seen as soon as the pair has formed. It is a short chase of fifteen to forty seconds' duration in which the male chases after the female while she seems to try to avoid him with an erratic flight path. The chase usually starts while the pair are feeding on their own, away from the nest site and other martins. The male may sing while chasing the female. Pair-chase seems to stop after the female starts laying eggs.

The other chase, rape-chase, involves several males and one female. This chase occurs primarily during nest-building, a period of about three to four weeks. The sight of the female gathering nesting material seems to stimulate other paired males to try to jump on her and mate with her while she is on the ground or to chase after her as she carries nesting material and try to mate with her when she lands. Usually there are two to six males chasing a single female. She may try to elude them by landing near her nest or diving into cover. Her mate often joins in, seeming to try to keep the other males away from her. Occasionally, fights occur between males as they vie for the female. Rape-chases are usually a little longer in duration than pair-chases, lasting up to a minute.

Nest-Building

Placement: In western states, martins will nest in any cavity, such as crevices in buildings, bridges, or rock cliffs, but prefer old woodpecker holes in trees and cactuses; in the Midwest and East, martins live almost exclusively in human-made nest boxes
Size: Cavity at least 6″ × 6″ × 6″ with an entrance hole 2½ inches in diameter
Materials: Grass stems, twigs, paper, mud, and green leaves

Nest-building begins about two to three weeks after the male and female have paired and about a month before egg-laying. The first attempts at nest-building are usually made by the male. You may see him picking up twigs, dropping them, flying halfway to the nest before dropping them, or even taking them to the nest but not getting them through the hole, so that they fall to the foot of the nest box.

The female, starting slightly after the male, does most of the real nest-building, with the male generally just following her on her trips to gather material. The first construction consists of grass stems and twigs. After continuing to work on this for several days she usually stops for one or two weeks, then resumes, usually in the morning hours, bringing twigs and, possibly, mud into the cavity.

The nest is a mat of material that slopes toward the back of the cavity and has a shallow cup to contain the eggs. Mud may also be used against the walls of the cavity.

Once the stick layer is complete, both birds begin to bring fresh green leaves in and lay them on top of the other materials. This usually starts just before egg-laying begins and continues right through the incubation phase, during which it is most frequent. One theory about the use of green leaves is that as they decay, the

leaves give off gases that may deter the presence of feather mites in the box. Another theory is that the moisture from the leaves keeps the eggs from drying out. Leaf-bringing stops after the young hatch.

Occasionally, if the male comes to the nest with material and the female is in the entrance, he may try to put the material in another hole. Also, the female may sometimes put some bits of material in an adjacent hole.

The entire nest-building period, from the male's first attempts to a few days before the beginning of egg-laying, lasts three to four weeks. During this period the female and male sleep together in the hole where the nest is being built.

Breeding

Eggs: 5 or 6. White
Incubation: 15–16 days, by female only
Nestling phase: 27–35 days, usually 28 days
Fledgling phase: About one week
Broods: 1

Egg-Laying and Incubation

During egg-laying the male tends to sing more. He also stops chasing the female and has fewer fights with neighboring males since territories are well established. Thus, the colony seems more peaceful at this time in the breeding cycle.

The female lays her eggs in the early morning, one each day; however, if there is a cold spell she may interrupt her laying for a day. Five or six eggs are the normal number, although first-year females tend to lay only four. If the eggs are destroyed or the young killed, the female may lay a replacement clutch about ten days later.

Incubation is done entirely by the female and starts after the laying of the last egg. When the female leaves the nest to feed, the male comes to the nest. He may remain in the nest, in the hole with his head out, or on the porch in front, until the female returns. On occasion he may leave before she returns. He does not have an incubation patch and so does not incubate the eggs even though he is occasionally on the nest.

The female is on the nest about 70 percent of the time during daylight hours. She also sleeps in the nest hole on the eggs every night; the male may or may not join her.

The incubation period is usually fifteen to sixteen days, but may be a day or two longer if the weather is cold.

Nestling Phase

The young hatch over a period of a day or two. The difference in their hatching time may be due to some partial preincubation development of the eggs that were laid earliest. For the first five days the female broods the young continuously while the male brings them food. Over the next five days, she broods them less during the day and instead helps bring them food. The number of trips to the nest with food is, on average, about ten per hour for both parents combined. At this early stage, fecal sacs are taken by the parents in their bills and dropped away from the nest.

By about the third week the nestlings have grown most of their body feathers, opened their eyes, and excrete their fecal matter out the nest entrance. In the fourth week they begin to grow their flight feathers.

The female broods the young every night until they are at least two weeks old. After this point she and the male may roost in other holes of their territory, or, if a pair have claimed only one nest hole, they might quietly slip into a neighbor's extra nest hole for the night at this time. They may also leave the colony site, going up to a half mile away, and roost in trees with other martins. These roosts may contain adults from several colonies.

The nestling phase usually lasts twenty-eight days but may vary from twenty-seven to thirty-five days. The young have to be good fliers when they leave the nest for there are generally no nearby perches for them to land on and they have to be able to return to the nest to roost. They are also sometimes attacked by other adults when they first leave. (*See* Fledgling Phase)

There are recorded cases of females entering a breeding colony late in the season, killing young in a nest hole, and taking over the hole and the mate and raising her own brood. This could be a strategy chosen by young females that arrive late and do not have mates or nest holes.

Fledgling Phase

When there is a platform or ledge outside a martin nest the young may come out up to four days before fledging and sit, exercise their wings, and be fed by the parents. If vagrant martins, especially subadult males, are in the area when the parents of these young are away, they may attack the young and try to pull them off the ledges. The young respond by retreating into the nest hole.

If other young fledglings of the same age show up on the ledge near the nest hole, the parents may feed them, apparently not able to tell them from their own young. Occasionally, more-advanced young wander into other nest holes, trampling the young there or keeping them from getting food and thereby leading to their death.

The young fledge over a period of up to three days. They usually take their initial flight in the first two hours of the morning, often following a parent away from the nest hole. Usually one fledgling leaves at a time, but occasionally several fly off at once. When a fledgling is flying away from the colony for the first time, other adults often fly after it, pecking and chasing it. The parent tries to drive them off.

Once each young bird leaves, it perches away from the nest box on trees, wires, or antennas. The young from a given brood are at first dispersed, but through various calls they gradually get together at some open perching area, up to a half mile away from the nest, where they are fed by the parents. At the end of the day, the fledglings follow the parents back to the colony and spend the night scattered in various nest holes. The next morning the fledglings sort themselves out and return to the same grouping areas they used the previous day, where they are again fed by the parents. This continues for two to three days. The two parents care equally for the young, who sit, sun, preen, and, occasionally, take short flights, returning to the perch immediately. The brood may also return to the nest before big storms.

Food transfer to the fledglings at the grouping areas is at first done by perched adults to perched young; then the adults hover and feed perched young; finally, the young fly out after the arriving adults and food is transferred in midair. The young birds generally flutter their wings before receiving food.

Certain subadult males, or adult males or females, will attack fledglings at grouping areas, and because of this they have been dubbed "raiders." They may land on the backs of the fledglings and peck at them or chase them. When the fledglings give their calls, their parents come and defend them against the raiders.

It has been suggested that harassment of fledglings by mature birds might lessen the chances of the young birds' returning to that site to breed in future years, thus reducing the competition for nest sites — an advantage to the raiders.

After two to three days at a grouping area, the family may leave and possibly go to areas where it is easier for the young to feed on their own. The young are probably independent after about a week or more. Parents return to the nest holes about a week or more after the young fledge and often engage in defense of one or more nest holes for another week or so. Then the birds gradually leave the nest area and join with other martins in feeding areas and communal roosts before and during their southward migrations.

Even though early writers, including Audubon, thought purple martins had three broods, it is now well established that they have only one, even in the South. Two broods occur only in unusually warm years, when the first brood can be started almost a month earlier than normal due to the prevalence of aerial insect life that the adults can feed on.

Young birds return to the general area in which they were born to start their own breeding the following year.

Plumage

DISTINGUISHING THE SEXES Males two or more years old are all a shiny blue-black with no lighter feathers showing. One-year-old males are like the female, which is duller on top and light-breasted, but they may have varying amounts of dark blue blotches on their breasts. The female is dull blue in back, with a grayish breast and, occasionally, a faint grayish collar over the nape of her neck.

DISTINGUISHING JUVENILES FROM ADULTS Juveniles look similar to adult females but have grayish foreheads, more so in the juvenile female than the juvenile male.

MOLTS Purple martins have one complete molt per year. This starts soon after breeding, in late July and early August. The adult birds begin their molt with their body feathers and some of their primaries — their longest wing feathers. Males generally start this molt slightly earlier than females. Before migration the birds have molted most of their primaries. During migration their molt is arrested.

Once the birds arrive on the wintering ground, the molt is resumed with loss of the rest of the primaries, the secondaries, the tail feathers, and completion of the head molt, which was started earlier. The molt is completed by February, before the birds migrate north.

The molt of juvenile birds is slightly different in that they do not start molting any flight feathers until they have completed migration.

Seasonal Movement

Spring migration for purple martins starts in late January, with the birds arriving at extreme southern breeding grounds in early February. The birds move north at the rate of about five degrees latitude every two weeks. By late April they have started to arrive at forty degrees latitude, a line running roughly from Philadelphia to San Francisco. Older males usually migrate earliest, followed by older females, who in turn are followed by first-year birds.

Temperature in the weeks before arrival has a definite effect on arrival date. If temperatures are above normal, the birds may arrive up to five days earlier, and if temperatures are below normal they arrive up to five days later. In the same area, certain colonies of martins may traditionally arrive earlier than others. This may be because these are larger colonies and there is more competition for the good nest sites. In general, the majority of birds arrive about a week after the first arrivals. It is not true that some birds arrive early as "scouts" and then go back and tell the other birds to follow.

Fall migration occurs in September and October. The birds travel to South America, where they spend the winter. Eastern birds seem to move gradually south during the day, with frequent stops for feeding. When they get to the Gulf Coast, many migrate over water in a continuous flight. They arrive in Central America and then continue south. Their winter range includes Brazil, Venezuela, and other adjacent areas. It is not fully clear how western populations migrate, but it is generally assumed that they move along the coast through Mexico and then Central America.

Flock Behavior

In fall, after breeding, purple martins in some areas of the country gather at night in huge roosts containing as many as one hundred thousand birds. The birds usually gather about an hour before dusk at pre-roosting sites, such as telephone wires, where they feed, bathe, and preen. Then just before dusk they all leave and enter a roosting area, which is usually a group of trees. The birds leave the roost up to several hours before daylight. Both the pre-roosting spot and roost spot may change in the course of a season.

In the West there are martin roosts that are used throughout the breeding season. One such roost was studied in Arizona. About three thousand birds started to use the roost upon arrival in spring. Through the incubation period females would stay on the nests and not return to the roost. When the young fledged they also joined the roost and swelled its number to as many as thirteen thousand birds. It was estimated that this roost served all martins within an area of five hundred and fifty square miles.

Martins are very socially oriented birds and often fly up as a group over a colony site to attack potential predators, such as hawks, owls, gulls, or even blue jays. Together they will chase them out of the area, swooping just above them but rarely hitting them. They may also chase off dogs or cats wandering into the area.

Bob Hines

Common Raven
Corvus corax

THE RAVEN'S LARGE SIZE, DEEP AND VARIED CALLS, AND ALMOST HAWKLIKE habits have inspired myths in many cultures about its prominence in the world of nature. And yet, despite all of this interest in the bird, its life and behavior are still shrouded in mystery. This may in part be due to the raven's preference for nesting in remote areas of the plains and mountains. In mountainous areas you often do not encounter the raven until you reach the barren wastes above tree line. In other areas, the favorite nesting spots of ravens are high among the crags of remote cliffs.

The raven is also extremely wary of humans, spotting you from almost as far as half a mile away as you approach a nest, and then flying up and calling at your approach. Because of this, studying ravens is best done through a scope or powerful binoculars when you are lucky enough to have found a nest.

As for the rest of the bird's behavior, you must take the bits and pieces you can see as the birds hunt or fly to and from their roosts. From these short observations you will begin to form a larger picture over the years. In many cases you may be able to learn new facts about the raven's life, in addition to those presented here.

BEHAVIOR CALENDAR

	TERRITORY	COURTSHIP	NEST-BUILDING	BREEDING	PLUMAGE	SEASONAL MOVEMENT	FLOCK BEHAVIOR
JANUARY							▓
FEBRUARY		▓	▓				▓
MARCH			▓	▓			▓
APRIL				▓		▓	▓
MAY				▓			▓
JUNE				▓			▓
JULY				▓	▓		▓
AUGUST				▓	▓		▓
SEPTEMBER				▓	▓	▓	▓
OCTOBER				▓	▓	▓	▓
NOVEMBER							▓
DECEMBER							▓

DISPLAY GUIDE

Visual Displays

Aerial-Tumbling

Male or Female   *Sp*

Elaborate flight maneuvers including steep dives, somersaults, birds flying near each other, and undulating flights.

CALL: None

CONTEXT: Done mostly in late winter and spring between members of a pair as they re-establish their bond near the nest site. *See* Courtship

Throat-Fluff

Male or Female   *Sp Su F W*

Bird fluffs out neck and head feathers, stands tall, and, in the case of the male, may walk about stiffly. In more intense versions, the bird assumes a horizontal posture and may repeatedly bow its head.

CALL: None, or male may give the Gro-call and the female may give squeaky or mechanical clacking calls

CONTEXT: Given by birds that are asserting their dominance in a hostile or sexual context. The less-dominant bird may crouch down and slightly spread its wings. *See* Flock Behavior

Allo-Preening

Male or Female   *Sp Su F W*

One bird runs its bill through the feathers on the head or throat of another bird.

CALL: None

CONTEXT: A bird that fluffs its head feathers is

often preened by its mate or another bird. *See* Courtship, Flock Behavior

Auditory Displays

Raven calls are complex and sometimes difficult to distinguish. The sound of calls varies among individuals, and individuals also have a strong tendency to incorporate other imitated sounds into their individual repertoire. The most common sounds are listed below; a few others are mentioned later in this chapter. Calls of the male are generally lower and louder than those of the female.

Krok-Call
Male or Female   *Sp Su F W*

The most common raven call — a short, low, clipped sound.

CONTEXT: Given during any perceived disturbance. Often given in response to your presence.

Gro-Call
Male or Female   *Sp Su F W*

A low, deep throaty note. Slightly longer and lower than the Krok-call.

CONTEXT: Given between members of a family, especially when food is being exchanged between parents and young or between mates, or when parents are approaching the nest with food. *See* Breeding

Crong-Call

Male or Female   *Sp*

A low, resonant call sounding like a deep bell.

CONTEXT: Context not known for sure. Has been heard from pairs as they circle over the nest in spring. Not known whether it occurs in other seasons.

Ruh-Call

Male or Female   *Sp Su F W*

A medium-pitched call that has an "uh" sound to it.

CONTEXT: Given loudly by fledglings as a way of announcing their location to their parents, and given more softly by subordinate birds when being confronted by a dominant bird.

Kra-Kra-Call

Male or Female   *W Sp Su*

A rapidly repeated series of short, harsh sounds. Higher pitched than other raven calls, sounding more like the caw of a crow.

CONTEXT: Given by young when begging for food and by the female as she receives food from her mate during mate-feeding. *See* Breeding, Courtship

BEHAVIOR DESCRIPTIONS

Territory

Type: Nesting
Size: Area right around nest
Main behavior: Aerial dives upon intruder
Duration of defense: Mostly in breeding season

Breeding pairs of ravens are territorial to the extent that they defend a small area around their nest against most other ravens and hawks. Defense usually consists of aerial dives upon the intruding bird. Ravens also have a more extensive range that overlaps with the ranges of other ravens and in which they are not aggressive. Pairs of ravens may roost on their territory each night or they may join a communal roost. *See* Flock Behavior

There are occasional reports of three or more ravens in the area of a nest or even at the nest during the breeding season. The third raven is most likely an immature bird from a previous brood, possibly taking part in raising the new brood.

There are also, occasionally, loose flocks of birds seen wandering about during the breeding season. It has been suggested that these are immature birds or nonbreeding birds without territories that roam about taking advantage of local food sources. *See* Flock Behavior

Courtship

Main behavior: Soaring, aerial acrobatics, Allo-preening
Duration: At the start of the breeding season

Once paired, ravens seem to stay together throughout the year and throughout their lives. When a member of a pair dies, a replacement mate is usually found within a very short time. Individual ravens can develop very characteristic habits and sounds that can enable you to recognize individuals and pairs from year to year.

At the start of the breeding season, the pair return to the nest area of the previous year. Their first visits may be brief, with subsequent visits being increasingly longer.

Three features of their behavior at this time may be part of courtship. One is close, coordinated soaring in circles over the nest site with their wingtips almost touching. In these maneuvers the male is usually above the female. Another feature is aerial acrobatics by the male, who either does steep sudden dives and then rises up or actually does tumbling in midair. A third behavior occurs when the birds are perched close to each other. They may touch bills several times and one bird may run its beak down through the breast feathers of the other in Allo-preening.

An additional behavior, called unison flight, can be seen at pre-roosting sites. These are highly coordinated flights distinguished by the birds' doing certain flight activities in unison. Their function is unknown but they look very much like courtship activities. *See* Flock Behavior

Nest-Building

Placement: Usually built in either a protected crevice of a ledge or cliff or the tops of trees, often evergreens; occasionally built on human structures
Size: Varies from 3 to 5 feet in diameter with an inner cup about 1 foot in diameter
Materials: Foundation of large sticks, earth, grass clumps; thinner twigs make up the rim of the nest; the cup is lined with bark shreds and any hair, wool, or similar fibers that the bird can gather

A raven nesting area may be used for up to a hundred years, spanning many generations of individual birds. Pairs that build in trees tend to continue this habit from year to year and the same is true of cliff nesters. Nest sites may be used two years in a row, especially cliff nests, but more often a pair seem to have two or more sites in an area that they shift between from year to year.

Both sexes build the nest, but in any given pair either the male or female may do the majority of the work. In one case nest-building was done by the female only, with the male accompanying her to and from the nest and sometimes doing aerial acrobatics during these trips. The pair usually give the Krok-call to each other while on these trips and near the nest. When at the nest the male may perch nearby while the female adds the material to the nest.

Completing the nest takes two to two and a half weeks. Sticks are generally broken off trees by the birds and not collected from the ground. The sticks may be up to three feet long and almost an inch in diameter. If the sticks fall from the nest, and they often do, the birds make no effort to pick them up and use them. When the birds work on the inner lining they may get in the nest and turn around, molding it with their body.

Since new twigs are collected each year, even when the birds are reusing an old nest, seeing the new twigs in the nest in spring may be an indication that a particular nest is occupied.

Breeding

Eggs: Average 3–5. Greenish with various amounts of brown blotches
Incubation: 18–19 days, by female only
Nestling phase: About 6 weeks
Fledgling phase: 5 to 6 months
Broods: 1

Egg-Laying and Incubation

There is usually a period, varying from a few days to four weeks, between the completion of the nest and the laying of the first egg. What determines the length of this wait is not known. The eggs are laid one each day until the clutch is complete. From the start of egg-laying, the female buries the eggs in the soft insulating material of the nest, probably to keep them warm during the cold temperatures of their early nesting season. She may also remain on the nest, but does not start actual incubation until the next-to-last or last egg is laid.

One clue to incubation's having started is seeing only one raven soaring above the nest. That would generally be the male since the incubating female spends most of her time on the eggs.

During incubation, the male brings food to the female and gives it to her at the nest or nearby. When the female is about to receive the food she may flutter her wings close to her body and give the Kra-kra-call. The Gro-call may also be heard from either bird around the time of food transfer.

When the male is not actively hunting for food for himself or the female, he is usually perched near the nest on a dead branch or ledge. The female occasionally leaves the nest, at which time the male may come to the nest, but he does not actually incubate the eggs.

If you approach a raven nest during the incubation phase, both birds are likely to fly up while you are still almost a half mile away and they will soar over the area giving the Krok-call.

If for some reason the eggs are destroyed, a replacement clutch is usually laid within three weeks, with the same number of eggs or one fewer than in the first clutch.

Nestling Phase

The young hatch over a period of a day or two. The female eats the shells and broods the young for about two and a half weeks. During this time the male does most of the feeding; after that, both parents participate in feeding the young. During the first weeks of intensive brooding, when the female takes periodic leaves, the male stays near or on the nest until she returns.

The parents at first bring small food items for the young, picking the items apart before offering them to the young. Later in the nestling phase larger food items are left at the nest and the young pull them apart to eat. The parents may also come to the nest with water in their crops, which is then fed to the young. On cold days the young are buried in the nest lining for warmth; on hot days the female may wet her underfeathers and cover the young to give them relief from the heat.

The young lift their tails as they defecate over the rim of the nest. The nest rim or cliff ledge can have large white stains resulting from this behavior; the stains may help you locate the nest.

The young may fledge over a period of several days.

Fledgling Phase

After taking their first flight, at which time they are already fairly good fliers, the fledglings return to the nest and continue to be fed in the vicinity for a week or more. After this, they move away to other areas but remain with their parents for five to six months after fledging. The whole family may move to areas providing better feeding opportunities.

Plumage

DISTINGUISHING THE SEXES The male and female raven look alike except that the male is generally larger than the female. The calls of the male are also lower than those of the female.

DISTINGUISHING JUVENILES FROM ADULTS The young have some brown on their wings and tail, whereas adults are solid black. The throat feathers of the young are not as noticeably long as those of adults.

MOLTS Adults have one complete molt per year, which occurs from July to October.

Seasonal Movement

Migration of ravens has not been well studied. Some pairs may remain in the area of their nest throughout the year, while others may migrate. Birds from extreme northern areas are the ones most likely to migrate. Fall migration occurs in September and October; people who monitor hawk migrations occasionally see small groups of ravens migrating at the same time. Where they go on migration is not known.

Flock Behavior

Large communal roosts of ravens occurring in fall and winter have been regularly reported. Once established, a roost site seems to be used for decades, or possibly even longer. One roost in Oregon was in low vegetation in a marsh. Although the roost often changed its specific location from night to night, it remained in the same general area all winter. It was slightly protected by a low ridge. This roost formed in mid-October and disbanded in mid-March. The most birds used the roost in January, when the roost population once reached over eight hundred birds. After January, the numbers decreased. This coincides with the beginning of some nesting activity.

In the Oregon roost, pre-roost gathering sites were openings in denser vegetation where the ravens gathered to feed. The birds fly to the pre-roost site via a direct route and in a style in which they alternate several flaps with a glide. They may fly singly or in groups of up to five birds. Birds that arrive at the pre-roost in pairs may then do coordinated flights in which they flap and glide in unison, do a roller-coaster action as a pair, or chase one another in fast flight, all done silently. Occasionally, two birds may hop off together, one crouching with wings partially opened and the other standing tall doing the Throat-fluff display. The birds might rub bills and then one might draw its bill down through the breast feathers of the other in Allo-preening. Up to four birds may posture together in this way.

The birds are conspicuous at pre-roost sites but then fly low and directly to roost sites, where they become inconspicuous. Birds leave for the roost site about a half hour before sunset and, when they reach the roost, dive down into it.

The seasons in which raven roosts form and their reason for forming need to be studied further. Sometimes roosts seem to form as a result of an abundance of food, such as when one occurs near a dump or where there is a large source of carrion. Roosts have also been recorded at all months of the year, with populations at each roost varying.

Most paired birds roost near their nest during the breeding season and, in many cases, throughout the year. Therefore, it is believed that many roosts may be composed of young or non-breeding individuals.

In some regions, there seem to be flocks of ravens that do not breed and that remain in areas where there are few nesting ravens. Whether these birds are all sexually immature is not known. The autumn and winter roosts of ravens that occur in some areas may be the result of many of these smaller flocks joining together each night.

In late summer, immature birds who are independent of their parents may flock with other ravens of similar-age, and possibly with other adult ravens that have not paired or have paired but, for some reason, do not have a territory. Within these flocks there are occasionally pairs of birds that stay closely associated with each other; their relationships are not known. These flocks may also form the basis for a communal roost.

It has been estimated that ravens may fly up to twenty-three miles to the roost.

Eastern Bluebird
Sialia sialis

FEW OTHER BIRDS HAVE CAPTURED THE HEARTS OF SO MANY, AND IT'S EASY
to see why; in terms of physical beauty and endearing tameness,
the Eastern bluebird has few rivals. Nearly everyone who has
bluebirds breeding on his or her property feels blessed, as if he or
she were specially chosen for the honor.

The bluebird was once a more common bird, but since the turn
of the century its populations have greatly declined. This may
have been due to the introduction from Europe, in 1851, of the
house sparrow and, in 1890, of the starling, two birds who are
aggressive competitors for the nest holes needed by the bluebird.
In addition, traditional bluebird nest sites, such as old apple or-
chards and tree holes along field edges, became more scarce as the
country continued to become more urban and suburban. Bluebirds
are also susceptible to cold weather and suffer great mortality
during very severe weather on their wintering grounds in the
southeastern United States.

Fortunately, Eastern bluebird populations are now on the
upswing, aided by the efforts of the North American Bluebird
Society and its inspirational founder, Dr. Lawrence Zeleny, in en-
couraging the construction, setting up, and monitoring of bluebird
nesting boxes all across the country. However, more nesting boxes
are still needed. This is an area where amateurs can still make a
valuable contribution toward the conservation of this beautiful
bird. For more information on setting up trails of bluebird boxes,
you can contact the North American Bluebird Society (see Appendix A).

BEHAVIOR CALENDAR

	TERRITORY	COURTSHIP	NEST-BUILDING	BREEDING	PLUMAGE	SEASONAL MOVEMENT	FLOCK BEHAVIOR
JANUARY							▓
FEBRUARY						▓	
MARCH	▓	▓				▓	
APRIL	▓			▓			
MAY	▓			▓			
JUNE	▓			▓			
JULY	▓			▓			
AUGUST					▓		
SEPTEMBER					▓		
OCTOBER						▓	
NOVEMBER						▓	
DECEMBER							▓

DISPLAY GUIDE

Visual Displays

Flight-Display
Male   *Sp Su*

Bluebirds vary their flight in several ways to communicate with one another. One Flight-display is a butterflylike flight in which wings are fluttered in slow deep wingbeats; another is hovering in front of the nest hole; and a third is a lopsided flight, with wings seeming to move out of synchrony and unevenly.

CALL: Song or Chip-call

CONTEXT: Given during territory formation and courtship. May function as territorial or mate advertisement. *See* Territory, Courtship

Wing-Wave
Male or Female   *Sp Su*

Bird lifts one or both wings at a moderate speed while perched. Wings may also be raised and quivered.

CALL: Song

CONTEXT: Done by male as part of courtship when female is near; may be done at nest entrance. May be done as greeting between pair. *See* Courtship

Wing-Flick
Male or Female   *Sp Su*

Wings are rapidly flicked and tail quickly spread at the same time.

CALL: Chip-call

CONTEXT: Given during aggressive encounters between bluebirds; often followed by a chase. *See* Territory

Head-Forward

Male or Female   *Sp Su F W*

Perched bird leans forward and assumes a horizontal posture with beak and head pointed forward.

CALL: Chip-call or none

CONTEXT: Occurs during aggressive encounters such as when other bluebirds approach the nest. May be combined with gaping bill or bill-snapping at the intruder.

Auditory Displays

Song

Male or Female   *Sp Su*

A series of six to eight soft, low-pitched, melodious whistles. May be strung together in a long series. Occasionally, just the first part of the song may be given. Although it is true that both sexes sing, the song of the male is louder, more brilliant, and heard more frequently.

CONTEXT: Used during territorial advertisement, courtship, and during some disturbances at the nest, such as your approach. *See* Territory, Courtship

Tur-a-Wee-Call

Male or Female   *Sp Su F W*

One to three soft, low-pitched, melodious whistles. Sounds like the written "tur-a-wee."

CONTEXT: Given as a location note between mates, family members, or flock members. Also given by parents as they approach the nest with food. *See* Courtship, Breeding

Chip-Call

Male or Female   *Sp Su F W*

A short, harsh call that may be given singly or in a rapid series making it sound like a chatter. CONTEXT: Given in moments of alarm, such as when an intruder comes near the nest.

BEHAVIOR DESCRIPTIONS

Territory

Type: Mating, nesting, feeding
Size: 2 – 25 acres
Main behavior: Song, Flight-display, chases
Duration: From arrival of male to end of breeding

Territories usually occur in open land with tree holes or nest boxes and some clear ground from which the birds can gather insects to eat. Size of territory varies, being generally larger at first and then shrinking with pressure from other arriving bluebirds and when more duties are required around the nest. It can be anywhere from two or three to twenty-five acres and may contain many nest holes, even though the birds will use only one. Bluebird pairs do not usually nest closer than about three hundred feet from each other, but sometimes they may nest closer.

Bluebirds may migrate or, in some cases, winter near their breeding ground. In either case they usually return to their previous breeding territories in early spring. In some cases, males and females arrive separately; in others, the pair arrive together. First-year bluebirds often return to breed near the area where they were hatched.

Bluebird behavior during the first few days on the territory may be erratic. They may show up for only a few hours in the morning, feed, visit nest holes, and then leave for the rest of the day. Gradually, they spend more and more time on the territory. Even after spending a few days visiting a nesting territory, a pair may fly off to a new area. Inclement weather may also cause bluebirds to vacate their territory temporarily.

The male claims the territory by giving Song from perches around the territory. He may also do Flight-displays accompanied by Song as he flies from one perch to another, or from a perch to the nest hole. Both male and female may defend the territory against the intrusion of other bluebirds, but each only defends against other bluebirds of its own sex. Thus, males will fight with males and females with females.

Aggressive interactions may include chases and several displays, such as Song, Chip-call, Wing-flick, and Head-forward. These can occur at a nest box or at a territorial border. Song duels between two males at territorial borders may be common where two pairs are nesting close together.

Other actions of bluebirds at this time that may function to advertise territory are short semi-circular flights, a straight rapid flight between two perches while continuously singing, and an action in which the male bluebird perches at the nest entrance with tail spread and repeatedly pokes his head in and out of the nest hole. This latter action is sometimes done with nest material in the bird's beak.

In the fall, bluebirds may inspect and be aggressive at nesting boxes. They may even add a little nesting material.

Courtship

Main behavior: Song, Flight-display, Wing-wave
Duration: For a few days after the arrival of the female on the territory

Before the female shows up on the territory, the male may do a great deal of Song, often singing as many as twenty times per minute. Once the female has arrived, the male tends to sing at a slower rate, more like five to ten times per minute.

When a female first arrives on the territory, or the pair arrive together, the male will start courtship activities with several displays. He will do Flight-display with Song, ending up at a prospective nest site. Here he will cling to the entrance or a nearby perch and do Wing-wave while continuing his singing. After this he may go to the nest hole and, while clinging to the entrance, rock back and forth, putting his head and shoulders in and out of the nest hole and looking around between each rocking motion. He may be carrying a bit of nest material while doing this. The male may also land on the nest box with his back to the female and spread his tail and droop his wings, exposing his vibrant blue back. He may lift and quiver his wings and pivot, appearing to dance.

These four elements are the main parts of early courtship: Song, Flight-display, Wing-wave, and the bird's poking his head in and out of the nest hole. They can be done in just about any order, and they can be repeated many times at the same or a different nest hole. Occasionally, the male may also fly at the female and chase her for a short distance.

At first the female seems to show little interest, but gradually she will approach a nest hole where a male has displayed and perch nearby or look in and out of the hole. Often the male will then enter the hole and may even sing softly while in the hole. The female usually leaves after a first visit, in which case the male continues his displays. She often approaches several boxes before she finally enters one. In the interludes between visits to nest holes she may also Wing-wave and give Song.

If she enters a nest hole while the male is also inside, this is a good sign that the two are paired. Following this she takes more of the lead around the nest site and is able to displace the male from any place that she flies to. She may be dominant over the male at this point.

Following this initial courtship phase the male and female will do fewer displays but will keep in contact with the soft Tur-a-wee-call and may do Wing-wave when coming together. Bluebirds also do courtship feeding, with the male bringing food to the female. Sometimes the female crouches down, flutters her wings, and makes peeping noises before the male feeds her. Courtship feeding continues throughout the early part of breeding.

Copulation in bluebirds may be seen starting before nest-building and continuing into the incubation phase. During copulation the male gets onto the back of the female, who is crouched down, and may even peck at her head.

Bluebirds are usually monogamous and male and female may remain together in successive seasons, especially if their previous breeding is successful. Occasionally, a bluebird finds a new mate for a second brood during a nesting season. There are also several recorded instances of polygamy. In one case, two males defended a territory together, mated with one female, and all three parents successfully raised two broods. In another case, a paired male, who had nestlings, courted and mated with another female and started a second brood. *See* Egg-Laying and Incubation

Nest-Building

Placement: In a natural tree hole or a nesting box. For information on constructing a nest box contact the North American Bluebird Society (see Appendix A)
Size: Inside diameter 2½–3 inches
Materials: Base of fine grasses, pine needles, weed stalks, and fine twigs; inner lining of finer grasses and, rarely, hair or feathers

Bluebirds build their nests in boxes or natural cavities in trees. They have also been known to use unusual nest locations, such as stacks of drain pipes, tin cans, cliff swallows' nests, cannons, and other cavities.

The nest is a loose structure of dried grasses and weeds with or without a distinct lining. When there is a lining it is made of finer grasses and, possibly, horsehair or feathers.

Although the male and female may enter several nest holes during courtship, it is the female that selects the final site. One to six weeks may go by between when the nest site is chosen and when the nest is actually built. A female may also start to build several nests and then choose one of these in which to lay her eggs.

The male may carry around bits of nesting material during

courtship and nest-building and, occasionally, may add some to a nest, but the actual building is done mostly by the female. Building may take from two to twelve days but is usually accomplished within four to five days. If there is cold weather, nest-building may be temporarily stopped.

Breeding

Eggs: Usually 4 or 5. Clear blue or, occasionally, white
Incubation: 12–18 days, average 13–14, by female only
Nestling phase: Usually 16–20 days
Fledgling phase: 3–4 weeks
Broods: 2 or 3

Egg-Laying and Incubation
After the completion of the nest, egg-laying may not begin for a week or more. Bluebirds usually lay one egg per day, with laying

usually occurring before midmorning. Incubation begins after the last egg is laid, but the female may spend brief periods sitting on the eggs prior to this.

Incubation is done by the female. The length of the periods spent attending the eggs varies with weather and other unknown factors. When the female leaves the nesting box for a break or to feed herself, the male may come to the box, occasionally going in but not actually settling on and incubating the eggs, since he has no brood patch. This has been verified by studies in which researchers had special blinds that enabled them to look into the nest box. At night, males often stay in the box with the female, standing next to her on the rim of the nest while she incubates the eggs. Throughout this period the male continues to bring some food to the female.

Another interesting behavior revealed in the course of studies in which researchers have been able to observe inside the nest box is called "tremble-thrust." In this, the female pokes her bill down through the nesting material and vibrates it. This occurs throughout the incubation and nestling phase. Why she does this is not known, but it may serve to shake parasites and other debris down to the base of the nest and away from the eggs and young.

Recent studies of bluebirds have shown that occasionally a female bluebird may lay an egg in another female bluebird's nest. This is called egg-dumping and, although it is known to occur among other birds (see Wood Duck), it is just beginning to be studied in songbirds. Through the examination of blood samples taken from bluebirds and their young without harming them, scientists have found that in 9 percent of bluebird nests the young are from more than one father or more than one mother. This could only occur if the female had mated with more than one male, or if another female had egg-dumped in her nest. The reasons for either one of these fascinating breeding strategies are still unknown.

Nestling Phase
The young hatch in the approximate order in which the eggs were laid. There can be one or more infertile eggs in a clutch. After the

eggs hatch the shells may be eaten or carried away. At first the young are practically "naked," covered only with scattered gray down. The female broods them for the first few days to keep them warm until they have developed more feathers and can regulate their own body temperature.

At first the male delivers most of the food, giving it to the female, who then feeds it to the young. Later, when the female no longer needs to brood the young as much, she participates in collecting food, and then both parents feed the food directly to the young. At first the young are fed soft insects, such as small caterpillars, but later they are fed larger adult insects, such as beetles and grasshoppers, and, sometimes, berries.

The young grow very fast, their eyes opening on the fourth to seventh day. The primary wing feathers are noticeable on the fourth day and tail feathers show on the eighth day. For the first week they give soft peeping sounds, which change to a harsher "zee" sound during the second week. By the twelfth day the young weigh almost as much as the adults and their down has been replaced by the blue and gray juvenal plumage. By the fifteenth day they are completely feathered.

In the early days of the nestling phase the parents may eat the fecal sacs. By the seventh or eighth day they carry them up to fifty yards away from the nest. Another clue to the stage of the nestling phase is how far the parents reach into the nest. In the early stages they go all the way into the box or cavity to feed the young; later they feed them while clinging to the entrance.

If one of the young dies during the nestling phase, the parents may remove it, unless it is too large for them to handle.

Fledgling Phase

When the young leave the nest varies. If they are disturbed during the late nestling phase they may leave prematurely and at great risk to their survival. Generally they leave when about seventeen to twenty days old. This is preceded by the adults' feeding them less until they leave the nest.

Within a period of about two hours the whole brood may leave the nest, although sometimes one or more members wait longer, even until the next day. On their first flight the nestlings are often capable of flying seventy-five to a hundred yards, often landing in the lower branches of trees and then working their way up to the higher branches. They usually start to give the Tur-a-wee-call as soon as they leave the nest and this may help the parents locate them during food trips. Both parents will usually continue to feed the young for three to four weeks or more. However, if the female starts in on another brood, the male will do all feeding of the fledglings.

The female bluebird may begin a new nest for a second or third brood in the same or a different nest box. She may begin her new brood as early as three to four days after her earlier brood has fledged. Occasionally, one or more of the young from a previous brood remain in the area and actually help feed the nestlings of the following brood. They are referred to as "helpers at the nest."

There are also reports of other adult bluebirds being helpers at the nest. In one case, a one-year-old male raised a brood and then in the same season returned to his parents and helped them feed their brood. It has also been reported that adults who are feeding their own fledglings have also fed other fledglings that wandered onto their territory.

Bluebirds that have nested successfully usually remain paired for future nestings. Following an unsuccessful nesting, one or both members of a pair may leave the nesting area and possibly nest elsewhere. Males have been known to move three miles to a new territory; a female traveled twelve miles after a nest failure. The cause of nest failures is most often bad weather or predation by raccoons, other mammals, snakes, house wrens, or house sparrows.

Plumage

DISTINGUISHING THE SEXES The male has bright blue plumage on back and head; female has grayish blue plumage on back and head.
DISTINGUISHING JUVENILES FROM ADULTS The juvenile's plumage is similar to the adult female's but with brownish spots all over its

breast. Juveniles have a partial molt in August and September, in which the spots are lost; they then resemble adults.

MOLTS Bluebirds have one complete molt per year, occurring in August and September.

Seasonal Movement

In the northern portion of their range, bluebirds migrate in winter to the southeastern United States and on down to Mexico and Central America. Migration occurs in October and November, and the birds generally travel in small or large flocks.

Occasionally, individual bluebirds do not migrate, but winter near where they previously bred, especially if there is sufficient food and their previous nesting attempt has been successful. Sometimes these wintering adults may remain with their offspring as a family group through the winter until the next breeding season.

Bluebirds in the southern portion of their range generally are year-round residents.

Flock Behavior

Bluebirds in winter often move about in small, loose flocks of five to ten birds, or occasionally more. They feed together on fruits and berries such as sumac and multiflora rose. Occasionally, these bluebirds come to feeders that offer raisins and suet mixtures. During cold weather the birds may roost together in nest boxes or other protected spots. In some cases the birds even arrange themselves side by side, with their heads pointing inward, possibly to conserve warmth.

Dark-Eyed Junco
Junco hyemalis

EVERY FALL WE AWAIT THE ARRIVAL OF THE "SNOW BIRDS" FROM THE north where they breed. The name comes from the junco's plumage, which has been described as "leaden skies above, snow below." This name more aptly describes the slate-colored form of the junco. Ornithologists used to think there were four separate species of juncos, white-winged, slate-colored, Oregon, and gray-headed. Now they are all considered one species, the dark-eyed junco. We tend to think of them as "snow birds" because we see them most when the snow is here.

Juncos are a favorite at winter bird-feeding stations throughout the United States and lower Canada. Much of the study of juncos has been of their winter flock behavior. There is still a lot to be learned about their courtship and breeding behavior.

Juncos tend to winter at the same spot each year and stay in fixed flocks with a stable dominance hierarchy. You can look for signs of this hierarchy at your winter feeder. More dominant birds will do Pecking-attacks, lunging at subordinates, who give way or avoid them. Occasionally, two birds will square off, face one another, and do the Head-dance display, in which they repeatedly throw their heads up and down. Sometimes things escalate into a fight, with the birds clawing at each other and rising up into the air, although generally things are settled by displays alone. As spring approaches, juncos start singing and do more Flight-pursuits, with the male chasing the female. Part of their courtship may take place in the flocks.

At night juncos often roost in the same place. It is fun to follow the flock from your feeder to see where they will roost. Usually it will be in some dense conifer where they will be protected from cold and predators.

BEHAVIOR CALENDAR

	TERRITORY	COURTSHIP	NEST-BUILDING	BREEDING	PLUMAGE	SEASONAL MOVEMENT	FLOCK BEHAVIOR
JANUARY							▓
FEBRUARY							▓
MARCH							▓
APRIL	▓	▓				▓	
MAY	▓			▓			
JUNE	▓			▓			
JULY				▓			
AUGUST					▓		
SEPTEMBER					▓		
OCTOBER						▓	▓
NOVEMBER							▓
DECEMBER							▓

DISPLAY GUIDE

Visual Displays

Pecking-Attack

Male or Female   *Sp Su F W*

With feathers sleeked, body horizontal, neck extended, and tail sometimes fanned, bird rushes toward another bird. If the other bird does not move, it may be pecked.

CALL: Kew-call, Zeet-call

CONTEXT: Done by more dominant bird during aggressive encounters.

Flight-Pursuit

Male   *Sp Su F W*

Short aerial chases in which one bird remains several feet behind the other bird. They may make abrupt turns. Tails are fanned.

CALL: Tsip-call, Zeet-call, Kew-call

CONTEXT: Given during aggressive encounters and as part of courtship.

Tail-Up

Male or Female   *Sp Su F W*

Tail is fanned and raised above the level of the head and neck is retracted. Bill may be open and wings slightly raised, making bird appear larger.

CALL: None

CONTEXT: Functions as a threat in aggressive encounters.

Head-Dance

Male or Female   *Sp Su F W*

Two birds face each other, extend necks, and raise and lower bills repeatedly.

CALL: Song, Kew-call, Zeet-call

CONTEXT: Given in aggressive or sexual encounters.

Auditory Displays

Song

Male   *Sp Su F W*

A musical trill that is variable. It may be all on one pitch or vary up or down in pitch, and can also vary in speed. Sometimes two or three trills on different pitches are strung together.

CONTEXT: Given during aggressive encounters or during courtship.

Kew-Call

Male or Female   *Sp Su F W*

A sharp musical sound. May be repeated.

CONTEXT: Given in aggressive encounters. Juncos that were feeding gave it as a rival approached, who then often stopped or retreated.

Zeet-Call

Male or Female   *Sp Su F W*

A short buzzing sound.

CONTEXT: Given in mildly aggressive situations. Often given when birds are arriving at a feeder or landing in a tree.

Tuck-Call
Male or Female   *Sp Su F W*
A harsh, short note.
CONTEXT: May express alarm. Given by a bird
that has become separated from the flock. A
longer distance contact note.

Tsip-Call
Male or Female   *Sp Su F W*
A high-pitched short note. Given singly or in
groups of three or more notes.
CONTEXT: This is the most frequently given
sound of juncos as they move about together.
Given from the ground or while in flight. May
function to keep flock members in contact.

BEHAVIOR DESCRIPTIONS

Territory

Type: Nesting, mating, feeding
Size: 2–3 acres
Main behavior: Song, chases
Duration of defense: Throughout breeding period

Male juncos arrive on their breeding grounds ahead of the females and begin singing from tall trees. Singing consists of musical
trills that may be quite variable. Males may have individually
unique song repertoires that can be distinguished from those of
other males. An intruder will be chased by the resident male.

Territory size is usually two to three acres but may be smaller.
Territories often include openings in the forest canopy in which
there is some sort of vertical outcrop or bank, since this is where
juncos prefer to nest, although they may also nest under bushes in
more open areas.

Courtship

Main behavior: Chases, Song
Duration: From arrival of female until incubation phase

When the female arrives, the male at first responds aggressively and chases her. She does not leave but persists in staying on the territory. Several writers have reported that the birds will display to each other by hopping with wings drooped and tail fanned, exposing the white outer tail feathers. A good sign that the two have paired is that they move about the territory together, usually staying within 50 feet of each other. The male also sings considerably less.

More study is needed of juncos' courtship behavior.

Nest-Building

Placement: Usually on the ground in a depression, often concealed by overhead vegetation
Size: Inner diameter 2¼ inches; interior depth 1¼ inches; outer depth 2½ inches
Materials: Dried grasses, moss, twigs, pine needles, bark, feathers, hair, fern rootlets

The nest-building is done primarily by the female, although the male may bring some material. Nests are usually in or near a vertical wall, such as a rocky outcrop, an exposed soil bank, or upturned tree roots, although they may also be nestled at the base of a tree, under some blueberry bushes, or under a brush pile. Many nests are concealed by overhanging vegetation such as ferns, moss, or branches. Rarely, nests may be built in a tree such as a conifer.

Eggs: 3–6, usually 3–5. Grayish or pale bluish with speckles and blotches of reddish browns and grays, often clustered at one end
Incubation: 12–13 days by female
Nestling phase: 9–13 days
Fledgling phase: 3 weeks
Broods: 1 or 2

Egg-Laying and Incubation

The female does all of the incubation. Eggs are laid one per day and incubation usually begins after the last egg is laid. In one study, when there were four eggs in a clutch, incubation was initiated after the third egg was laid. The first three eggs hatched at the same time and the last egg hatched a day later.

This study also found that three-egg clutches were most common in a renesting after a successful brood was raised, or after a nesting failure.

Nestling Phase

When the young hatch, both parents feed them. The average frequency of feeding is eight times per hour. Food for the nestlings consists mainly of soft insects and some seeds. When large insects are brought, the wings and legs are carefully removed before the insects are given to the young.

Both parents remove fecal sacs, eating them for the first few days, then carrying them away thereafter. There are several reports of the parents' carrying the fecal sacs to a specific tree limb and wiping them off there. In one case, a telephone wire was used, which gleamed white for quite a distance, until rain washed it off.

Brooding is done only by the female. The young open their eyes on the second day, and the feathers of the juvenal plumage are unsheathed by the seventh day. If the nest is approached at or before this point, the young will show fear by hunching down in the nest. After eight days the young become progressively more restless and jostle about in the nest. If the nest is disturbed after this point the young may leave prematurely.

Fledgling Phase

After the young leave the nest they may be fed by the parents for three weeks or more. In one brood that was studied, the male was seen feeding the young twenty-four days after they had left the nest; on the twenty-seventh day, however, he became aggressive and chased a begging young. This male was seen coming to a feeder with his young forty-six days after they had left the nest, but he did not feed them.

If there is a second brood, the female may begin nest-building as early as two days after the first young have left the nest. The male then takes over the care of the fledglings.

There is one unusual instance recorded of a male who took over the care of a deceased male's brood. On the morning the egg hatched (only one out of three eggs was fertile), the father of the brood was found dead. By that afternoon, the female had acquired a new mate. The female alone fed the young and several copulations between this pair were observed over the next three days. By the fourth day, the new male was also feeding the nestling. Both parents successfully raised this young.

Plumage

Junco plumage varies geographically, which is the reason ornithologists formerly classified juncos into four separate species. All the forms are now considered one species, the dark-eyed junco. It is now accepted that five forms of this one species may be distinguished in the field: the Slate-colored junco (in the East), the Oregon junco (in the West), the Pink-sided form of the Oregon junco (in the northern Rocky Mountains), the White-winged junco (in the Black Hills of Wyoming and South Dakota), and the Gray-headed junco (in the southern Rocky Mountains). Since there may be interbreeding, intermediates between these distinct forms are sometimes seen.

DISTINGUISHING THE SEXES In the Slate-colored form, it is difficult to determine junco age and sex reliably in the field. Usually the very darkest birds are males. Males are also usually the darkest in the White-winged form, which looks similar to the Slate-colored form but has white wing bars and more white in the tail. In the Oregon form, the male has a black hood, whereas the female has a grayer hood. In the Pink-sided and Gray-headed forms, the sexes are alike.

DISTINGUISHING JUVENILES FROM ADULTS Juveniles of all races of the species have streaked breasts, whereas adults have solid breasts. MOLTS Juncos molt once a year, in August and September.

Seasonal Movement

Juncos leave the northern portion of their breeding range during October and November and migrate in flocks to their wintering areas in lower Canada and throughout the United States. Juncos usually return to the same wintering areas every year. Generally, the first arriving birds are older and more dominant; they are then joined by younger birds. Males tend to winter farther north than females, so the proportion of males in a winter flock will be higher the farther north it is.

Flock Behavior

Winter flock behavior has been the most-studied aspect of junco life. Juncos form fixed flocks on their winter grounds. The flock forages in a defined area of about ten to twelve acres. Not all the members of a flock always move about together, so you may see varying numbers of juncos, but they all stay within the defined foraging area. They do not defend this area from other juncos, though the areas are mutually exclusive. Occasionally, some juncos may temporarily wander into another flock's area, particularly if food is scarce.

The flock has a stable social hierarchy that is organized into a straight-line pecking order. Males tend to dominate females and adults dominate immature birds.

Much of the behavior seen at feeders is the result of the flock hierarchy. Juncos express dominance by several displays, including Pecking-attacks, Head-dance, and Flight-pursuits. Flight-pursuits are done by males to females, and since this display is done more frequently in spring before the flocks migrate, it may serve some courtship function.

When juncos feed, they seem to space themselves out over an area. Pecking-attacks are frequently seen and given more by dominant birds. Dominant birds seem to interact more with other dominant birds, while subordinate birds tend to avoid dominants. Occasionally, two juncos will hold their ground; the birds will face one another and do Head-dance. A chase may follow, or sometimes the birds actually fight, clawing at each other and rising several feet in the air, although this is rare.

During the day, flocks often follow a routine feeding circuit. At night they roost together and may repeatedly use the same spot. Much chasing among the branches and flashing of the tail feathers can be seen as they settle down for the night. The flashing tail feathers may actually help the flock locate one another and stay together.

As spring approaches, there is an increase in male singing and, as noted, Flight-pursuits. By March and April, the birds have left to return to their breeding grounds.

tists are continually looking for explanations for this occurrence; the answers at the moment are only tentative.

Even though much study has been done, there are many basics of white-throat behavior that need to be looked at more carefully. These include the functions of their various calls, nest-building behavior, and courtship. Although white-throats are wary of humans during the breeding season, they are still good subjects for behavior-watching, for their territories are small, male and female have differently colored heads, and you can come to recognize the Songs of individual males in a given area.

BEHAVIOR CALENDAR

	TERRITORY	COURTSHIP	NEST-BUILDING	BREEDING	PLUMAGE	SEASONAL MOVEMENT	FLOCK BEHAVIOR
JANUARY							■
FEBRUARY							■
MARCH					■	■	
APRIL	■				■	■	
MAY	■	■	■	■			
JUNE	■			■			
JULY	■			■			
AUGUST					■		
SEPTEMBER							
OCTOBER						■	
NOVEMBER						■	
DECEMBER							■

DISPLAY GUIDE

Visual Displays

Tail-Flick
Male or Female   *Sp Su F W*
Tail is rapidly and repeatedly flicked up.
CALL: Pink-call
CONTEXT: Given during moments of alarm or excitement.

Crest-Raise
Male or Female   *Sp Su F W*
The feathers on the top of the head are erected, giving the bird a slightly crested appearance.
CALL: Pink-call or none
CONTEXT: May occur during aggressive interactions or during moments of alarm. May be accompanied by Tail-flick.

Flutter-Flight
Male   *Sp Su*
The bird flies in a slow, fluttering manner. Flight covers only a few yards and is low over the ground.
CALL: None
CONTEXT: Done by the male as he approaches the female for copulation. Usually done in response to the Trill-call of the female. *See* Courtship

Auditory Displays

Song
Male or Female   *Sp Su F W*

A beautiful series of whistled notes on two or three pitches. Very distinct from other white-throat calls. Individual birds can be recognized by the distinctness of their Song.

CONTEXT: Given in territorial advertisement and courtship. Also given during migration and on wintering grounds, where it may also be territorial. Given primarily by males, but also by "white-striped" females (see Plumage). *See* Territory, Courtship, Seasonal Movement, Flock Behavior

Tseet-Call
Male or Female   *Sp Su F W*

A drawn-out, high-pitched call with a distinct "eee" sound.

CONTEXT: Given between members of a flock while feeding, on migration, or between mates on the breeding ground. May be a contact call, helping birds know each other's whereabouts. *See* Flock Behavior

Pink-Call
Male or Female   *Sp Su F W*

A very short, high-pitched call. Sounds like "pink" or, when quieter, "tip."

CONTEXT: Quieter versions may serve as contact notes; louder versions may be given in moments of alarm, such as disturbances near the nest. *See* Territory, Breeding

Chup-Up-Call

Male or Female   *Sp Su F W*

A rapid series of low-pitched sounds, like "chup-up-up-up-up."

CONTEXT: May be given during aggressive interactions between birds, such as territorial interactions between males. *See* Territory

Trill-Call

Male or Female   *Sp Su*

A high-pitched trill lasting for a second or more.

CONTEXT: Most often heard from the female prior to copulation or in response to male Song. *See* Courtship

BEHAVIOR DESCRIPTIONS

Territory

Type: Mating, nesting, feeding
Size: ½ to 2½ acres
Main behavior: Song, chases
Duration of defense: From arrival on breeding ground into fledgling phase

White-throated sparrows tend to breed in or at the edge of clearings, especially in areas where there are scattered shrubs and small trees. The males arrive on the breeding ground ahead of the females and seem to return to the territories of previous years. Soon after arriving, they start their territorial defense.

When the male first arrives, much of his territorial behavior consists of chases. These chases may be accompanied by Crest-raise, the Chup-up-call, and the Pink-call. After a day or so the male begins to spend more time singing from several prominent perches within his territory. He usually does not perch on the very top of shrubs and trees, but down slightly from the top.

Studies of white-throats show that they seem to avoid singing while a neighboring white-throat is singing. This results in what sounds like duets, or trios, with each bird giving Song in turn. By avoiding overlap with another male's Song, a bird maximizes the usefulness of its own Song.

The female soon arrives, and both birds remain in the territory through their breeding phase. Territories can be quite small, ranging from less than half an acre to about two and a half acres. The size of the defended territory may get smaller as the season progresses, especially in areas where several pairs are trying to nest at once.

In midsummer, when the young fledglings become independent, territorial boundaries are no longer defended as strictly. The male, however, tends to remain on the territory even though the female and young may have wandered away.

Song is also heard from white-throated sparrows outside the breeding season, especially during migration, when the birds stop over in areas to feed. It is possible that Song in this context may serve to advertise temporarily claimed feeding areas of each bird — in a sense mini-territories defended for a short time.

Courtship

Main behavior: Song, chases
Duration: For the week after the arrival of the female on the breeding ground

Females arrive on the breeding grounds one to two weeks after the males. Females that have bred in previous years tend to return to the same breeding ground each year.

Once the female arrives, singing by the male decreases. But it may increase again if the first nesting fails and the two renest.

Not much is known about courtship in white-throated sparrows. Females may at first be attracted by singing males. This may be followed by some initial chases of females by males, which are followed in turn by a gradual acceptance of each bird by the other. After this, male and female move about the territory in fairly close association, possibly keeping in contact with each other through the Tseet-call or Pink-call.

During this time the male is extremely wary of neighboring males that come near his mate. He may even try to adjust his territorial borders if the female shows a preference for part of a neighbor's territory. This heightened defense of the female and the territory will reoccur if the first nesting fails and the pair renest.

About a week before egg-laying starts, the female may start to give the Trill-call. She may just give it on her own, or she may give it seemingly in response to the male's Song. In either case, it seems to signal her readiness for copulation. After hearing this call, the male may fly to her with Flutter-flight. She may quiver her wings and continue the Trill-call as the male mates with her.

Another feature of white-throated sparrow courtship has to do with their polymorphism (see Plumage). White-striped birds only pair with tan-striped birds. Therefore, in any pair there is always a white-striped member and a tan-striped member. Once you figure out which bird is male and which is female, you can easily recognize them through their striping for the rest of the breeding period.

Nest-Building

Placement: On the ground or slightly above, usually near the base of a shrub or small tree that is in or at the edge of a clearing
Size: Inside diameter 2½ inches, inside depth 1½ inches
Materials: An outer structure of coarse grasses and an inner lining of fine grasses and, possibly, rootlets and/or animal hair

There are very few recorded observations of white-throated sparrows building nests. The female seems to do most of the nest-building and this occurs primarily in the morning. New nests built from year to year or within the same season are not built in the same spot, but are usually quite removed from the previous nest location.

Breeding

Eggs: 4 – 6. Glossy, whitish to pale blue or pale green with dark markings
Incubation: 11 – 14 days, by female only
Nestling phase: 7 – 12 days
Fledgling phase: 3 – 4 weeks
Broods: 1, occasionally 2

Egg-Laying and Incubation
The eggs are laid one each day until the clutch is complete. Egg-laying occurs in the morning. Incubation starts after the laying of the last egg and is done only by the female. Incubation lasts for eleven to fourteen days.

During the incubation phase the male may stay away from the nest. He will occasionally give a quiet version of Song while foraging on the ground. This may function to keep the female aware of his location. The female leaves the nest to feed several times a day and the male may accompany her at these times.

White-throated sparrows are very secretive around the nest. The female is not easily flushed; she may stay still on the nest until you are within two or three feet of it. Even then, she may not immediately fly up, but, instead, will run a short distance along the ground before flying up. Calls, such as the Pink-call, are likely to be given by both male and female for as long as you are in the area of the nest. These may be accompanied by Tail-flicks. But the birds will still remain hidden, even when scolding an intruder.

Nestling Phase

White-throated sparrows have a fairly short nestling phase, lasting on the average eight to nine days. Since the white-throat is a ground nester, it is probably safer for the young to leave the nest as soon as possible, as they may be more subject to predation there. The parents are secretive during this stage as well; when the male or female brings food to the nestlings, it may not fly directly to the nest but, rather, may land farther away and either walk or hop toward it.

The young leave the nest before they can fly, hopping to nearby perches.

Fledgling Phase

Very little is known about this phase of white-throated sparrow life. Both male and female continue to feed the young, which seem to be able to fly within a few days of leaving the nest. The parents are quick to defend the young at this time by perching more visibly and giving Pink-calls.

This phase lasts three to four weeks. It seems that if the female and young move off the territory, the male may not join them; instead, he may stay on the territory until migration time.

Plumage

DISTINGUISHING THE SEXES There is no way to distinguish male from female white-throated sparrows by plumage. However, in terms of behavior, it is primarily the male that gives Song and defends the territory and only the female that incubates.

DISTINGUISHING JUVENILES FROM ADULTS Juveniles have grayish throat patches, some fine streaking on the breast, and may lack the yellow lores of the adult. Adults have bright white throat patches, yellow lores, and a clear breast.

MOLTS White-throated sparrows have two molts per year. A complete molt of all feathers occurs in late July and August. During this molt, most white-striped birds acquire buff-colored head stripes.

A partial molt of head, throat, breast, and flank feathers occurs in late March to early April. During this molt, the white-striped birds again acquire their bright white head stripes.

POLYMORPHISM There are two forms of white-throated sparrows: those with black and white stripes on their heads and those with brown and tan stripes on their heads. A bird is either one or the other form throughout its life. Both males and females can be either form. During breeding, tan-striped birds mate only with white-striped birds, an occurrence extremely rare among birds.

Seasonal Movement

White-throats winter throughout eastern North America and in southern portions of central and western North America. They migrate at night in medium-sized flocks that stop over during the day and feed in brushy areas. Song may be heard on their northward and, occasionally, on their southward journeys. Its function at these times is not known.

Banding records tend to indicate that females and young birds migrate farther south than adult males.

Spring migration occurs in April.

Flock Behavior

During migration and on the wintering grounds, white-throated sparrows tend to remain in small flocks of about five to fifteen birds, feeding on the ground, often in the company of other species, such as juncos, which also feed on the ground. Song is occasionally heard in winter, but much more common is the Tseet-call, which is given almost constantly as the birds forage together. It probably functions as a contact call, enabling the flock to stay together as it feeds.

Bobolink
Dolichonyx oryzivorus

AN OUTSTANDING FEATURE OF BOBOLINK-WATCHING IS WITNESSING THE courtship and territorial behavior of the males. While pouring out their Song, which sounds a little like several small melodious flutes playing ascending scales together, they do remarkable visual displays that heighten the effect of their wing and body colors.

Bobolinks tend to nest in loose colonies in open hay fields. This adds to the excitement of behavior-watching since there are lots of interactions between birds, many examples of displays, and discrete small territories. On the other hand, there can be so much behavior going on in a bobolink breeding area that trying to follow individuals gets confusing.

One of the interesting features of their behavior is the tendency of the males to be polygamous. Some males in a group may have up to four females nesting in their territories. These are most likely older males, with more experience and higher-quality territories. Much study of their interactions with their various mates has been done, and evidence suggests that they help raise the young in each brood as time permits, giving most of their help to their first mate of the season and her young. When these young are well on their way the male may shift his energy over to helping raise his later mates' broods.

Bobolinks are quite specific in their breeding-habitat needs. Open hay fields are a must, and as farming in some regions of the country diminishes so do populations of bobolinks. Where colonies of bobolinks have traditionally bred, it is important to preserve their habitat with regular mowing practices. Unfortunately, the right time to mow a field for hay is often just when the young are fledging. Careful observation of the behavior of a colony and delaying of mowing until one or two weeks after fledging, a time when the young can fly fairly well, will keep a colony producing and ensure its survival. We hope that reading this behavior summary will help you enjoy and preserve bobolinks better in your area.

BEHAVIOR CALENDAR

	TERRITORY	COURTSHIP	NEST-BUILDING	BREEDING	PLUMAGE	SEASONAL MOVEMENT	FLOCK BEHAVIOR
JANUARY					■		
FEBRUARY					■		
MARCH						■	
APRIL						■	
MAY	■	■	■	■		■	
JUNE	■	■		■			
JULY				■	■	■	■
AUGUST					■	■	■
SEPTEMBER						■	
OCTOBER							
NOVEMBER							
DECEMBER							

DISPLAY GUIDE

Visual Displays

Songspread

Male   *Sp Su*

Bird lowers head, spreads tail and wings, and fluffs nape feathers. May turn from side to side with a hopping motion. Done from perches or on the ground. Done at varying intensities, from full display described above to just nape ruffling.

CALL: Song

CONTEXT: This is the most common display during territory establishment. Low-intensity version used as territorial advertisement; high-intensity version done as threat gesture to other males. Also used to attract female. *See* Territory, Courtship

Songflight

Male   *Sp Su*

Bird flies slowly but with rapid, shallow wingbeats below the body plane. May be done high in the air, or low over the ground.

CALL: Song

CONTEXT: A circular version done low over the ground is usually associated with courtship of the female. A high flight over the territory helps establish territory ownership. Two males may do Songflight along a common border. During incubation several males in a given field may all rise up and do Songflight together. *See* Territory, Courtship

Bill-Flip

Male or Female   *Sp Su*

Bird quickly flips its bill up in the air two or three times in succession.

CALL: None

CONTEXT: Done mostly when two males are walking along a common territorial border or are perched near each other. Only rarely done by females. *See* Territory

Tail-Wing-Flick

Male or Female   *Sp Su*

Tail and wings are given a quick flick at the same time.

CALL: Male — Check-call; female — Quipt-call

CONTEXT: Given when intruders are near the nest. *See* Breeding

Wing-Lift

Male   *Sp Su*

Wings are lifted above body and held there for several seconds. Body feathers may be ruffled and tail spread.

CALL: Song

CONTEXT: Done by male during courtship phase when a female flies overhead or when a female is perched nearby. *See* Courtship

Dropping-Down

Male   *Sp Su*

After a low circular Songflight the male drops to a perch with legs dangling down and wings held over his back in a V. Wings may remain held up for several seconds after landing.

CALL: Buzz-call

CONTEXT: Done by male near female during courtship phase following a low circular Songflight. *See* Courtship

Auditory Displays

Song
Male   *Sp Su*
A complex but beautiful sound consisting of several distinct notes followed by a series of ascending warbles. Easily distinguished from all other bobolink sounds by its length and complexity. May be given in a long (four-to-six-second) or short (two-to-three-second) version. Other versions consist of song fragments or compound song, which is several songs strung together.
CONTEXT: Accompanies Songspread and Songflight displays. *See* Territory, Courtship
NOTE: Bobolinks give a wide variety of short calls. These are best distinguished by their *context* and by noting the *sex* of the bird. Naming and describing their different sounds are not particularly helpful in distinguishing them.

Zeep-Call
Female   *Sp Su*
A note with a "zeep" sound, repeated three to ten times in an evenly spaced pattern.
CONTEXT: Given by females early in the nesting season and often during conflicts with other females.

Quipt-Call
Female   *Su*

A short call like its written sound, accompanying Tail-wing-flick or hovering flight over the nest site. Given repeatedly.

CONTEXT: Given by the female while perched or in flight whenever there is potential danger near the nest or young. Given in response to human intruders. *See* Breeding

Seeyew-Call
Male   *Su*

A clear, descending whistle.

CONTEXT: Given by male as he hovers over intruders that are near nest or young during nestling and fledgling phases. *See* Breeding

Buzz-Call
Male   *Sp Su*

A short buzz repeated two or three times.

CONTEXT: Given by male during courtship as he does the Dropping-down display. *See* Courtship

Check-Call
Male   *Sp Su*

A rapidly repeated short harsh sound, given in a series of up to fifteen notes.

CONTEXT: Given by males during territorial conflict chases or accompanying Tail-wing flick display. *See* Territory, Breeding

Chunk-Call
Male or Female   *Sp*

A short, harsh note given once every three to four seconds.

CONTEXT: Used by male or female during disturbances by intruders early in the breeding season, such as before and during incubation. *See* Breeding

Pink-Call
Male or Female   *Sp Su F*
A short, metallic sound.
CONTEXT: Given after the breeding season by adults and immatures when they are gathered into flocks. This is the typical bobolink call during migration. *See* Breeding, Seasonal Movement

Fledgling-Calls
The sound of fledglings during the week or two before they flock with adults has been aptly described as the sound of a "stretched rubber band when plucked." This is replaced by the Pink-call when the birds join into flocks with adults.

BEHAVIOR DESCRIPTIONS

Territory

Type: Mating, nesting, feeding
Size: 1 acre on land with abundant food, 5 – 9 acres on land with sparse food
Main behavior: Song, chases, parallel walks
Duration: From arrival on breeding ground until fledgling stage

When males first arrive on the breeding ground they tend to feed together, move about the breeding fields, and do little displaying. After a few days, sometimes up to a week, they each begin to occupy separate areas where they spend more time doing the

Songspread and Songflight displays, both of which are accompanied by Song. Two neighboring males may alternately do Songspread. This is done from perches around their territory, such as atop small shrubs or, particularly, tall weed stems.

Intruding males are chased in flight for up to a minute and often high in the air. These intruders are often transient males, unaware of territorial boundaries and being chased from territory to territory by each successive owner. The chased bird may give the Check-call.

Boundaries between neighboring birds are established in two ways. One is through a version of Songflight where the two neighboring birds fly parallel to each other along a border and then circle back into their respective territories. Another way is through a behavior that occurs on the ground and has been called parallel-walk. Two neighboring birds land along a common border and walk or hop parallel to each other along it. The parallel-walk usually includes many other actions and displays. It may be interspersed with foraging by each male, or the males may stop to do displays such as Songspread or Bill-flipping, or they may take short runs along the border. Other displays done during parallel-walks include two actions of the head — one in which the head is turned to the side and another in which it is nodded downward. Both gestures tend to expose the golden nape of the male and hide its bill.

Parallel-walks are one of the first signs of territory formation and may continue for two to three hours at a time. A male may do parallel-walk with several different neighboring males in a day, or with the same one repeatedly over several days.

Occasionally aerial flights occur in which both males fly vertically up and, while facing each other in midair, grapple with their feet and peck at each other. Sometimes they lock feet and fall to the ground.

At first the males do not spend the whole day on their territories; they may go elsewhere to feed. Territorial activity is most intense in early to mid-morning. The central territories in a field area are occupied first, often by older, dominant males. Later-arriving males, often younger males, form territories on the periphery of these. As with many birds, territory size tends to shrink as more males arrive and there is more pressure and conflict.

Most territories are settled by the time females arrive, and then there is less Songspread and Songflight. During incubation Songflight is often seen done by all the males in a given area at once. As one rises in flight all the others join in.

Territory defense slows when the young hatch, for the males assist in feeding them and have less time. At the fledgling stage the young scatter into many different territories, and territories are no longer defended.

Bobolinks are often aggressive toward other species that try to nest in the same fields. They will chase after red-winged blackbirds, tree swallows, and other birds that fly into the area.

Bobolinks have strong site fidelity and usually return to within about fifty meters of where they bred in previous years. First-year males do not all get territories.

Courtship

Main behavior: Songflight, Wing-lift, Dropping-down
Duration: From arrival of the female to fledgling stage

Females arrive about one week after the males. Bobolinks are frequently polygamous, males pairing with up to three (or, rarely, four) females at the same time. Generally, all older territorial males in an area get one mate before any gets two. First-arriving females tend to be older (at least two years old) than later-arriving ones. Males that are polygamous get their second females several days after their first arrives. Sometimes early-arriving females that have mated are aggressive to later-arriving females that want to pair with the same mate.

One of the first displays a male does after seeing a female fly overhead or perch near him is Wing-lift accompanied by a short version of Song. Following this he may do Songflight.

Songflight during courtship takes a slightly different form from that during territory formation. It is generally done low, less than three meters off the ground, and it is circular and slower, with the male landing near where he started.

Following Songflight, the male may do Dropping-down, in which he drops down into the vegetation with legs dangling and wings held up over his back in a V. He gives the Buzz-call during this display and may hold his wings in a V for several seconds after landing. He may then crawl through the vegetation with wings partially spread as the female perches nearby. This whole sequence may be repeated every few minutes during intense courtship.

Another aspect of courtship is the chasing of the female by the male. When the female lands conspicuously and near the male, he often initiates a chase, with her flying erratically ahead while he dives at her from behind, but rarely hits her. Occasionally, several males will briefly join in the chase of one female. Chases are less frequent once egg-laying starts.

Once the two are paired the female will stay on the male's territory most of the time, only occasionally leaving to feed in communal feeding areas, and when she does this the male may accompany her.

Nest-Building

Placement: On the ground in open fields
Size: Inside diameter 2½ inches, inside depth 1 – 2 inches
Materials: Grass and sedge leaves, coarse on the outside with a fine lining

Bobolinks prefer to nest in open hay fields and they tend to nest near other bobolinks.

The nests are built at the base of grasses or wildflowers in a shallow depression either found or made by the bird. The nest tends to be next to clumps of thicker vegetation and often on the east or south side of that vegetation. The nest is not covered over but exposed to the sky.

The female does all nest-building and tends to collect materials at a distance from the nest, flying to and from the nest low over the ground, making her nest-building activities inconspicuous. The nest takes two or more days to complete.

Breeding

Eggs: 5 or 6, later nests 4 or 5. Cinnamon colored and heavily and irregularly blotched with brown
Incubation: 12 days, by female only
Nestling phase: 10–11 days
Fledgling phase: Three or more weeks
Broods: 1

Egg-Laying and Incubation

Egg-laying starts about two days after the nest is completed. One egg is laid each day until the clutch is complete. Females that arrive early lay five or six eggs in a clutch while those that arrive slightly later may lay only four or five. This may reflect the fact that the early arrivals are the older, more experienced females.

Incubation starts on the day before the last egg is laid and is all done by the female. She averages about twenty minutes on the nest and ten minutes off. While off the nest, she generally feeds in the territory and is sometimes accompanied by the male. Occasionally, the male and female leave the territory to feed in communal feeding areas nearby.

Your presence near the nest will probably cause the female or male to give the Chunk-call, or the male may fly up and hover over you and give the Seeyew-call.

Nestling Phase

The young hatch over a thirty-six-hour period and the female carries away the remains of the shells. In a colony of bobolinks, most hatching is quite synchronous and is completed within a single week.

For the first four days the young are brooded by the parents. When both parents are attending the nest, one may remain brooding until the other arrives with food, thus creating continuous brooding. In cases in which only the female is tending the nest she must divide her time between brooding and collecting food.

Both male and female feed the young, the young's diet consisting mostly of caterpillars. Males with more than one female generally help only the first female feed and brood, but they may adjust this habit depending on the condition of nestlings in each of their females' nests. After the young in the first nest get to be a certain age a polygamous male may switch his help to his second nesting female. At what stage he does this may depend on the availability of food and, again, on the condition of the nestlings in each nest. In some cases, males visit the second nest with food, as if to check on the young, before going to feed the young in the first nest. During the nestling phase the parents spend an increasing amount of time foraging in communal feeding areas.

During the first few days, fecal sacs either are not carried away or are too small to be seen. Later, they are carried away from the nest and dropped fifty or more feet away.

After four days, the young are no longer brooded and the adults just stay at the nest long enough to drop off the food. This is when the birds are most conspicuous around the nest and thus the best time to locate nests. However, bobolinks are very wary and will not approach the nest with food unless you are quite far away. Also, they may enter the grasses and walk to the nest. The best clue to nest location is the point from which the parents leave after feeding the young; this spot tends to be nearer the nest.

During the nestling phase the parents react strongly to your presence near the nest. They will do the Tail-wing-flick accompanied by the Check-call, or the male will hover overhead giving the Seeyew-call. The female may fly about in an excited manner giving the Quipt-call during this and the following stage. This makes it very obvious to the observer that he or she is near the nest or young.

The young stay in the nest for about eleven days, but will leave a day or two earlier if disturbed. This may be to their disadvantage, so it is best not to disturb the young or approach the nest late in the nestling phase.

There have been some instances of extra adults being seen with food at nests during the nestling phase — in some cases extra males and in others extra females. The relationship of these birds to the parents is not known. This needs further study.

Fledgling Phase

Within a few minutes of leaving the nest, the young fledglings begin giving their "buzzy" call note to help their parents locate them with food. When they first leave the nest they cannot fly, but within two days they can fly short distances; within five days they can fly well and fly after their parents to beg for food.

Fledglings eventually move far from the nest and within a day can be seen more than one hundred fifty feet away. This means that they scatter over other males' territories, precipitating the breakdown of territorial boundaries. Within a week of fledging, young and adults gather into a loose flock that grows in size as more young fledge. At this stage the young seem to beg for food from any adult (and it is not known whether adults recognize their own young). This flock stays together, moves about together, and flushes as a flock. When flying, the adults and immatures now give the Pink-call. The flock remains intact until it migrates.

Rarely, second broods are attempted by females that started nesting early and who have an abundance of food to utilize. But there are no documented cases of these females ever successfully raising young.

Plumage

DISTINGUISHING THE SEXES Male and female bobolinks are easily distinguished during the breeding season. Males have a black head, belly, and wings, with a buff-gold nape and white patches on the back. The female is buff colored all over with dark streaks on her back, wings, and sides. In winter plumage, male and female look similar.

DISTINGUISHING JUVENILES FROM ADULTS The juveniles look much like the female in her breeding-season plumage except that they lack streaks on the side and have yellower underparts.

MOLTS Bobolinks have two complete molts per year. A complete molt starts after breeding in late July. The female does not change colors, but the male changes completely and comes to look like the female, with even his bill color changing from black to a clay color. A second complete molt occurs in late winter before spring migration. The female retains the same coloration, but the male gets a black bill and his distinctive plumage.

Seasonal Movement

When breeding is finished, flocks of bobolinks move to marshy areas and cultivated fields where they feed on wild grains and roost in large numbers during the night. This is also when they undergo their molt. In late July some migration begins. Migration takes place primarily at night with the birds feeding during the day or stopping for several days to feed at particularly good spots.

Birds in western states and provinces move East and join up

with eastern birds migrating along the southeastern coast, ending up in Florida. Flights then take them across the Gulf of Mexico to Cuba, where they may stop and feed, and then across the Caribbean to South America. There, their migration continues into southern South America (southern Brazil, northern Argentina, Paraguay) where they winter mostly in grain fields and similar areas.

In spring they fly north by similar routes, across the Caribbean and the Gulf of Mexico, landing by mid-April on the coast from Florida to Louisiana. Again, they fly at night. The birds all fly north and then western birds fly west as well. The birds arrive in northern areas in early May.

A characteristic call during the flight of migration is the Pink-call.

Bobolinks' round-trip flight takes them over eleven thousand miles, the longest migration of any North American songbird (passerine).

Flock Behavior

Bobolinks join into flocks after the breeding season and roost and feed together as they undergo their molt and prepare for migration.

Appendix A: Bird-Conservation Organizations

LISTED HERE ARE ORGANIZATIONS THAT PROMOTE AND PROTECT VARIOUS species of birds. If you would like to learn more about and/or help support their work, write to these addresses.

General

International Council for Bird Preservation
United States Section, Inc.
801 Pennsylvania Ave. SE
Washington, DC 20003

Bluebird

North American Bluebird Society
Box 6295
Silver Spring, MD 20906

Common Loon

North American Loon Fund
Main St.
Humiston Building
Meredith, NH 03253
603-279-5000

Common Tern

Colonial Waterbird Society
c/o Archives—National Museum of Natural History
Smithsonian Institution
Washington, DC 20560

Hawks

Hawk Migration Association of North America
377 Loomis St.
Southwick, MA 01077

Hawk Mountain Sanctuary Association
Rte. 2
Kempton, PA 19529
215-756-6961

Institute for Wildlife Research
1412 16th St. NW
Washington, DC 20036

Raptor Research Foundation
12805 St. Croix Trail
Hastings, MN 55033

Osprey

The International Osprey Foundation, Inc.
289 Southwinds
Sanibel, FL 33957

Peregrine Falcon

The Peregrine Fund, Inc.
World Center for Birds of Prey
5666 West Flying Hawk Ln.
Boise, ID 83709
208-362-3716

Pheasants

World Pheasant Association
752 Swede Gulch Rd.
Golden, CO 80401

Purple Martin

Nature Society
Purple Martin Junction
Griggsville, IL 62340
217-833-2323

Purple Martin Conservation Association
c/o Edinboro University of Pennsylvania
Institute for Research and Community Services
Edinboro, PA 16444

Appendix B: Hawk-Watching

ONE OF THE MORE EXCITING EVENTS IN BIRD-WATCHING IS GOING TO special spots in fall or spring to see migrating hawks. At some of these locations it is possible to see literally thousands of hawks fly overhead in a single day. This is because hawks migrate by day and are often concentrated into large groups at certain points due to land formations, weather conditions, and the hawks' own preferences.

Hawk-watching during migration has been going on for decades, but it has recently gotten national attention and is becoming a widespread sport among bird-watchers and hawk enthusiasts. All of the hawks in this volume of *A Guide to Bird Behavior* can be seen on migration if you know where, when, and how to look. We have provided this appendix to help get you started hawk-watching and to give you the information you need to go further.

First of all, familiarize yourself with the information under Seasonal Movement for each hawk. There you will find a description of the bird's migration, clues on how to recognize the hawk during migration, and remarks on some of the features of its behavior during migration.

Following the Seasonal Movement information for each hawk is a Fall Migration Chart. This chart has arrows showing the major routes (when known) of the hawk's migration and dots showing hypothetical routes. In most cases there are lots of dots on the maps since there is still so much to be learned about actual migration routes.

Also shown on each map are several of the major hawk-watching sites, where hawks are regularly counted and careful records of their numbers are kept. Next to each site, the average number of the species seen migrating in a given fall is noted.

At the base of the map are listed the dates of peak migration and the average number of that species seen on a good day's observation.

These charts have been done only for fall migration, for several reasons. Fall migration is the most easily seen, the most spectacular, and the best-studied migration season. The spring migration of hawks is still a bit of a mystery. The numbers seen are never as large as in fall and the migration routes are not as concentrated. Even so, we have added some spring migration data at the bottom right-hand side of each chart. This includes the average spring count at two spring sites and the dates for peak spring migration. These spring sites are identified on the map by reference numbers.

All of this is only a guideline. There are hundreds of other hawk-watching locations and many other species seen on migration that are not covered here. No one can tell you all about hawk-watching, for there is still much to be learned and probably always will be. This is part of what makes it so exciting.

For more information on hawk migration and places nearby where you can go to watch for hawks, contact the Hawk Migration Association of North America (HMANA). After this you might contact one of the major hawk-watching sites nearest you for further information on places to see hawks and people who can help you see them. The addresses of some of these are also listed below.

Hawk Migration Association of North America
377 Loomis St.
Southwick, MA 01077

Some major hawk-watching sites:

Cape May Bird Observatory
P.O. Box 3
707 East Lake Dr.
Cape May Point, NJ 08212
609-884-2736

Derby Hill Bird Observatory
Sage Creek Rd.
Mexico, NY 13114

Hawk Mountain Sanctuary Association
Rte. 2
Kempton, PA 19529
215-756-6961

Hawk Ridge Nature Reserve
Duluth Audubon Society
c/o Biology Dept.
University of Minnesota at Duluth
Duluth, MN 55812

Holiday Beach Migration Observatory
1120 Clair St.
Ann Arbor, MI 48103

Whitefish Point Bird Observatory
Paradise, MI 49768

Raptor Migration Observatory
Golden Gate National Park Association
Fort Mason, Building 204
San Francisco, CA 94123–9970

Glossary

Bill-wipe — The act of wiping the bill across a branch during confrontations

Brood — The birds hatched from one clutch of eggs

Brooding — The act of sitting over the newly hatched young to keep them warm; done by the parents in the first few days of the nestling phase

Brood patch — A spot on the breast of an incubating bird where there are fewer feathers and an increased blood supply to help keep eggs warm during incubation

Call — An auditory display, generally simpler in structure than song

Clutch — A set of eggs for a single brood

Courtship — All behavior that involves the relationship between males and females in breeding condition

Display — A stereotyped movement or sound that a bird makes, and that, when used in certain situations, affects the behavior of other animals near the displaying bird

Fecal sac — A small mass of excrement, surrounded by a coating of mucus, which is excreted by a nestling and carried off or eaten by the parents

Fledgling — A young bird that has left the nest but is still dependent on its parents for some or all of its food

Home range — An area inhabited by a bird, but not necessarily defended against its own or other species

Incubation — The act of covering the eggs to keep them warm and further their development

Incubation patch — An area of the breast in which blood vessels increase to provide added warmth to eggs

Mate-feeding — The feeding of one adult member of a pair by the other; usually occurs only during the breeding season (sometimes called "courtship feeding")

Nestling — A hatched bird that remains in the nest and is cared for by the parents or other adults

Pair formation — The aspect of courtship that involves the pair's first encounters and their becoming committed to each other

Primary roost — A fixed location at which birds habitually gather during the inactive phase of their day

Range — An area regularly inhabited by a bird, but not consistently defended

Seasonal movement — Predictable large-scale movement of populations over the course of the year

Secondary roost — Like the primary roost, but used for a shorter period of time and during the active phase of the birds' day

Song — A complex auditory display that may be partially inherited and partially learned

Territory — Any defended area

Bibliography

THIS BIBLIOGRAPHY INCLUDES ONLY THE MAJOR WORKS THAT WERE USED IN writing this guide. Many other articles were read but not included here either because they were only marginally helpful or because their information was included in more recent updates. The amount of study that has been done by researchers varies widely from species to species and is reflected generally in the number and length of the studies listed for each bird.

COMMON LOON

McIntyre, J. W. 1978. Wintering behavior of common loons. *Auk* 95: 396–403.

———. 1983. Pre-migratory behavior of common loons on the autumn staging grounds. *Wilson Bull.* 95: 121–125.

Munro, J. A. 1945. Observations on the loon in the Cariboo Parklands, British Columbia. *Auk* 62: 339–344.

Olsen, S. T., and W. H. Marshall. 1952. The common loon in Minnesota. Minn. Mus. Nat. Hist. Occ. Pap. No. 5.

Palmer, R. S. 1962. *Handbook of North American birds.* Volume 1. New Haven: Yale University Press.

Powers, K. D., and J. Cherry. 1983. Loon migrations off the coast of northeastern United States. *Wilson Bull.* 95: 125–132.

Ream, C. H. 1976. Loon productivity, human disturbance, and pesticide residues in northern Minnesota. *Wilson Bull.* 88: 427–432.

Rummel, L., and C. Goetzinger. 1975. The communication of intraspecific aggression in the common loon. *Auk* 92: 333–346.

———. 1978. Aggressive display in the common loon. *Auk* 95: 183–186.

Sjolander, S., and G. Agren. 1972. Reproductive behavior of the common loon. *Wilson Bull.* 84: 296–308.

Southern, W. E. 1961. Copulatory behavior of the common loon. *Wilson Bull.* 73: 280.

Sutcliffe, S. 1982. Prolonged incubation behavior in common loons. *Wilson Bull.* 94: 361–362.

Tate, D. J., and J. Tate, Jr. 1970. Mating behavior of the common loon. *Auk* 87: 125–130.

Yeates, G. K. 1950. Field notes on the nesting habits of the great northern diver. *Brit. Birds* 43: 5–8.

GREAT BLUE HERON

Cottrille, W. P., and B. D. Cottrille. 1958. Great blue heron: Behavior at the nest. *Misc. Publ. Mus. of Zool.*, no. 102. University of Michigan.

Forbes, L. S. 1984. Extreme aggression in great blue herons. *Wilson Bull.* 96: 318–319.

Gibbs, J. P., et al. 1987. Determinants of great blue heron colony distribution in coastal Maine. *Auk* 104: 38–47.

Kernes, J. M., and J. F. Howe. 1967. Factors determining great blue heron rookery movement. *J. Minnesota Acad. Sci.* 3: 80–82.

McAloney, K. 1973. The breeding biology of the great blue heron on Tobacco Island, Nova Scotia. *Can. Field-Nat.* 87: 137–140.

Meyerriecks, A. J. 1960. Comparative breeding behavior of four species of North American herons. *Publ. Nuttall Ornith. Club*, no. 2.

Palmer, R. S. 1962. *Handbook of North American birds*. Volume 1. New Haven: Yale Univ. Press.

Pierce, P. A. 1982. Behavior of fledgling great blue herons in a Michigan rookery. *Jack-Pine Warbler* 60: 5–13.

Pratt, H.M. 1970. Breeding biology of great blue herons and common egrets in central California. *Condor* 72: 407–416.

Pratt, H. M., and D. W. Winkler. 1985. Clutch size, timing of laying, and colony reproductive success in a colony of great blue herons and great egrets. *Auk* 102: 49–63.

Quinney, T. E. 1982. Growth, diet, and mortality of nestling great blue herons. *Wilson Bull.* 94: 571–577.

WOOD DUCK

Armbruster, J. S. 1982. Wood duck displays and pairing chronology. *Auk* 99: 116–122.

Clawson, R. L., G. W. Hartman, and L. H. Fredrickson. 1979. Dump nesting in a Missouri wood duck population. *J. Wildl. Manage.* 43: 347–355.

Heusmann, H. W., and R. H. Bellville. 1982. Wood duck research in Massachusetts. Mass. Div. Fisheries and Wildlife. Research bulletin 19.

Johnsgard, P. A. 1965. *Handbook of waterfowl behavior.* Ithaca, NY: Cornell Univ. Press.

Korschgen, C. E., and L. H. Fredrickson. 1976. Comparative displays of yearling and adult wood ducks. *Auk* 93: 793–807.

Leopold, F. 1951. A study of nesting wood ducks in Iowa. *Condor* 53: 209–220.

Palmer, R. S. 1976. *Handbook of North American birds.* Volume 3. New Haven and London: Yale University Press.

AMERICAN WOODCOCK

Gregg, L. E., and J. B. Hale. 1977. Woodcock nesting habitat in northern Wisconsin. *Auk* 94: 489–493.

Krohn, W. B. 1971. Some patterns of woodcock activities on Maine summer fields. *Wilson Bull.* 83: 396–407.

Marshall, W. H. 1982. Does the woodcock bob or rock—and why? *Auk* 99: 791–792.

Morgenweck, R. O. 1984. Observations on postures and movements of non-breeding woodcock. *Wilson Bull.* 96: 720–723.

Nero, R. W. 1977. The American woodcock in Manitoba. *Blue Jay* 35: 240–256.

Rabe, D. L., H. H. Prince, and D. L. Beaver. 1983. Feeding-site selection and foraging strategies of American woodcock. *Auk* 100: 711–716.

Roberts, T. H. 1980. Sexual development during winter in male American woodcock. *Auk* 97: 879–881.

Samuel, D. E., and D. R. Beightol. 1973. The vocal repertoire of male American woodcock. *Auk* 90: 906–909.

Shissler, B. P., and D. E. Samuel. 1983. Observations of male woodcock on singing grounds. *Wilson Bull.* 95: 655–656.

Thomas, D. W., and T. G. Dilworth. 1980. Variation in peent calls of American woodcock. *Condor* 82: 345–347.

Worth, C. B. 1976. Body-bobbing woodcocks. *Auk* 93: 374–375.

COMMON TERN

Bannerman, D. A. 1962. *The birds of the British Isles.* Volume 11. London: Oliver and Boyd.

Cooper, D. M., H. Hays, and C. Pessino. 1970. Breeding of the common and roseate terns on Great Gull Island. *Proc. Linnaean Soc.* 73: 82–103.

Coulter, M. C. 1980. Stones: an important incubation stimulus for gulls and terns. *Auk* 97: 898–899.

Cramp, S., ed. 1985. *Birds of the western Palearctic.* Volume 4. Oxford: Oxford Univ. Press.

Hays, H. 1984. Common terns raise young from successive broods. *Auk* 101: 274–280.

Morris, R. D. 1976. Factors influencing desertion of colony sites by common terns. *Can. Field-Nat.* 90: 137–143.

Nisbet, I. C. T. 1973. Courtship-feeding, egg-size and breeding success in common terns. *Nature* 241: 141–142.

———. 1975. Selective effects of predation in a tern colony. *Condor* 77: 221–226.

———. 1976. Early stages of postfledgling dispersal of common terns. *Bird-Banding* 47: 163–164.

———. 1983a. Defecation behavior of territorial and non-territorial common terns. *Auk* 100: 1001–1002.

—————. 1983b. Territorial feeding by common terns. *Col. Waterbirds* 6: 64–70.

Palmer, R. S. 1941. A behavior study of the common tern. *Proc. Boston Soc. Nat. Hist.* 42: 1–119.

BALD EAGLE

Bortolotti, G. R., J. M. Gerrard, P. N. Gerrard, and D. W. A. Whitfield. 1983. Minimizing investigator-induced disturbance to nesting bald eagles. In *Proc. Bald Eagle Days* 1983, 85–103. Winnipeg, Manitoba.

Broley, C. L. 1947. Migration and nesting of Florida bald eagles. *Wilson Bull.* 59: 1–20.

Fraser, J. D., et al. 1983. Three adult bald eagles at an active nest. *Rap. Res.* 17: 29–30.

Gerrard, J. M. 1983. Bald eagle migration through southern Saskatchewan and Manitoba and North Dakota. *Blue Jay* 41: 146–154.

Gerrard, P., et al. 1974. Post-fledgling movements of juvenile bald eagles. *Blue Jay* 32: 218–226.

Gerrard, P. N., S. N. Wiemeyer, and J. M. Gerrard. 1979. Some observations on the behavior of captive bald eagles before and during incubation. *Rap. Res.* 13: 57–64.

Griffen, C. R. 1981. Interactive behavior among bald eagles wintering in north-central Missouri. *Wilson Bull.* 93: 259–264.

Hensel, R. J., and W. A. Troyer. 1964. Nesting studies of the bald eagle in Alaska. *Condor* 66: 282–286.

Herrick, F. H. 1924. Nests and nesting habits of the American bald eagle. *Auk* 41: 213–231.

—————. 1924. The daily life of the American eagle: late phase. *Auk* 41: 389–422 and 517–541.

—————. 1932. Daily life of the American eagle: early phase. *Auk* 49: 307–323.

—————. 1933. Daily life of the American eagle: early phase. *Auk* 50: 35–53.

Ingram. T. N. 1965. Wintering bald eagles at Guttenberg, Iowa. *Iowa Bird Life* 35: 66–78.

Knight, S. K., and R. L. Knight. 1983. Aspects of food finding by wintering bald eagles. *Auk* 100: 477–484.

Russock, H. I. 1979. Observations on the behavior of wintering bald eagles. *Rap. Res.* 13: 112–115.

Sherrod, S. K., C. M. White, and F. S. L. Williamson. 1976. Biology of the bald eagle on Amchitka Island, Alaska. *Living Bird* 15: 143–182.

Stalmaster, M. V. 1987. *The bald eagle.* New York: Universe Books.

Weekes, F. M. 1975. Behavior of a young bald eagle at a southern Ontario nest. *Can. Field-Nat.* 89: 35–40.

SHARP-SHINNED HAWK

Bowles, J. H. 1930. Nesting of the sharp-shinned hawk. *Murrelet* 11: 13–14.

Brown, W. J. 1916. The sharp-shinned hawk. *Ottawa Nat.* 30: 97–100.

Fischer, D. L. 1984. Successful breeding of a pair of sharp-shinned hawks in immature plumage. *Rap. Res.* 18: 155–156.

Ganier, A. F. 1923. Nesting of the sharp-shinned hawk. *Wilson Bull.* 35: 41–43.

Hart, C. G. 1934. Sharp-shinning in 1934. *Oologist* 51: 94–99.

Kerlinger, P., J. D. Cherry, and K. D. Powers. 1983. Records of migrant hawks from the North Atlantic Ocean. *Auk* 100: 488–490.

Millsap, B. A., and J. R. Zook. 1973. Effects of weather on Accipiter migration in southern Nevada. *Rap. Res.* 17: 43–55.

Mueller, H., D. D. Berger, and G. Allez. 1979. Age and sex differences in size of sharp-shinned hawks. *Bird-Banding* 50: 34–44.

Mueller, H. C., and D. D. Berger. 1967. Fall migration of sharp-shinned hawks. *Wilson Bull.* 79: 397–415.

Platt, J. B. 1976. Sharp-shinned hawk nesting and nest site selection in Utah. *Condor* 78: 102–103.

Reynolds, R. T., and H. M. Wight. 1978. Distribution, density, and productivity of Accipiter hawks breeding in Oregon. *Wilson Bull.* 90: 182–196.

Rust, H. J. 1914. Some notes on the nesting of the sharp-shinned hawk. *Condor* 16: 14–24.

Simpson, R. B. 1911. The sharp-shinned hawk. *Oologist* 28: 54–56.

Stabler, H. B. 1891. Nesting of the sharp-shinned hawk. *Oologist* 8: 161–162.

Willard, B. G. 1911. Nesting of the sharp-shinned hawk. *Oologist* 28: 61–63.

NORTHERN GOSHAWK

Bond, R. M. 1940. A goshawk nest in the upper Sonoran life-zone. *Condor* 42: 100–103.

———. 1942. Development of young goshawks. *Wilson Bull.* 54: 81–88.

Cramp, S., ed. 1980. *Birds of the western Palearctic*. Volume 2. Oxford: Oxford Univ. Press.

Dixon, J. B., and R. E. Dixon. 1938. Nesting of the western goshawk in California. *Condor* 40: 3–11.

Gromme, O. J. 1935. The goshawk nesting in Wisconsin. *Auk* 52: 15–20.

Lee, J. A. 1980. Survival of the smallest nestling in goshawks. *Rap. Res.* 14: 70–73.

Mueller, H. C., D. D. Berger, and G. Allez. 1977. The periodic invasions of goshawks. *Auk* 94: 652–663.

Parratt, L. 1959. Observations at a goshawk nest in northwestern Montana. *Wilson Bull.* 71: 194–197.

Schnell, J. H. 1958. Nesting behavior and food habits of goshawks in the Sierra Nevada of California. *Condor* 60: 377–403.

Sutton, G. M. 1925. Notes on the nesting of the goshawk in Potter County, Pennsylvania. *Wilson Bull.* 37: 192–199.

Zirrer, F. 1947. The goshawk. *Passenger Pigeon* 9: 79–94.

BROAD-WINGED HAWK

Banks, J. W. 1884. Nesting of the broad-winged hawk. *Auk* 1: 95–96.

Burns, F. L. 1911. A monograph on the broad-winged hawk. *Wilson Bull.* 23: 139–320.

Currie, J. D. 1901. Nesting of the broad-winged hawk. *Oologist* 18: 5–8.

Fitch, H. S. 1974. Observations on the food and nesting of the broad-winged hawk in northeastern Kansas. *Condor* 76: 331–333.

Heintzelman, D. S. 1975. *Autumn hawk flights.* New Brunswick, NJ: Rutgers Univ. Press.

Keran, D. 1978. Nest site selection by the broad-winged hawk in north central Minnesota and Wisconsin. *Rap. Res.* 12: 15–20.

Matray, P. F. 1974. Broad-winged hawk nesting and ecology. *Auk* 91: 307–324.

Mosher, J. A., and P. F. Matray. 1974. Size dimorphism: a factor in energy savings for broad-winged hawks. *Auk* 91: 325–341.

Mueller, H. C. 1970. Courtship and copulation in hand-reared BWHA. *Auk* 87: 580.

Mueller, H. C., and D. D. Berger. 1965. A summer movement of broad-winged hawks. *Wilson Bull.* 77: 83–84.

Riley, J. H. 1902. Notes on the habits of the broad-winged hawk in the vicinity of Washington, D.C. *Osprey* 6: 21–23.

Rosenfield, R. N. 1982. Unusual nest sanitation by a broad-winged hawk. *Wilson Bull.* 94: 365–366.

———. 1982. Sprig collection by a broad-winged hawk. *Rap. Res.* 16: 63.

———. 1984. Nesting biology of broad-winged hawks in Wisconsin. *Rap. Res.* 18: 6–9.

Rusch, D. H., and P. D. Doerr. 1972. Broad-winged hawk nesting and food habits. *Auk* 89: 139–145.

Smith, N. G. 1980. Hawk and vulture migrations in the neotropics. In Keast, A., and E. U. Morton, eds., *Migrant birds in the neotropics.* Washington, D.C.: Smithsonian Press.

Titus, K., and J. A. Mosher. 1981. Nest-site habitat selected by woodland hawks in central Appalachians. *Auk* 98: 270–281.

———. 1982. The influence of seasonality and selected weather variables on autumn migration of three species of hawks through the central Appalachians. *Wilson Bull.* 94: 176–184.

RED-TAILED HAWK

Ballam, J. M. 1984. The use of soaring by the red-tailed hawk. *Auk* 101: 519–524.

Fitch, H. S., F. Swenson, and D. F. Tillotson. 1946. Behavior and food habits of the red-tailed hawk. *Condor* 48: 205–237.

Gates, J. M. 1972. Red-tailed hawk populations and ecology in east-central Wisconsin. *Wilson Bull.* 84: 421–433.

Janes, S. W. 1984. Fidelity to breeding territory in a population of red-tailed hawks. *Condor* 86: 200–203.

Luttich, S. N., L. B. Keith, and J. D. Stephenson. 1971. Population dynamics of the red-tailed hawk at Rochester, Alberta. *Auk* 88: 75–87.

Mader, W. J. 1978. Comparative nesting study of red-tailed hawks and Harris' hawks in southern Arizona. *Auk* 95: 327–337.

Orians, G., and F. Kuhlman. 1956. Red-tailed hawk and horned owl populations in Wisconsin. *Condor* 58: 371–385.

Santana, E., R. L. Knight, and S. A. Temple. Parental care at a red-tailed hawk nest tended by three adults. *Condor* 88: 109–110.

Stinson, C. H. 1980. Weather-dependent foraging success and sibling aggression in red-tailed hawks in central Washington. *Condor* 82: 76–80.

Valentine, A. E. 1978. The successful nesting of a red-tailed hawk in an urban subdivision. *Jack-Pine Warbler* 56: 209–210.

OSPREY

Cramp, S., ed. 1980. *Birds of the western Palearctic*. Volume 2. Oxford: Oxford Univ. Press.

Green, R. 1976. Breeding behavior of ospreys in Scotland. *Ibis* 118: 475–490.

Grover, K. E. 1984. Male-dominated incubation in ospreys. *Condor* 86: 489.

Henny, C. J., and W. T. Van Velzen. 1972. Migration patterns and wintering localities of American ospreys. *J. Wildl. Manage.* 36: 1133–1141.

Ingram, C. 1959. The importance of juvenile cannibalism in the breeding biology of certain birds of prey. *Auk* 76: 218–226.

Jamieson, I., N. Seymour, and R. P. Bancroft. 1982. Time and activity budgets of osprey nesting in northeastern Nova Scotia. *Condor* 84: 439–441.

Poole, A. 1979. Sibling aggression among nestling ospreys in Florida Bay. *Auk* 96: 415–417.

———. 1982. Breeding ospreys feed fledglings that are not their own. *Auk* 99: 781–784.

———. 1985. Courtship feeding and osprey reproduction. *Auk* 102: 479–492.

Prevost, Y. 1979. Osprey-bald eagle interactions at a common foraging site. *Auk* 96: 413–414.

Stinson, C. H. 1977. Familial longevity in ospreys. *Bird-Banding* 48: 72–73.

PEREGRINE FALCON

Beebe, F. L. 1960. The marine peregrines of the northwest Pacific coast. *Condor* 62: 145–189.

Bird, D. M., and Y. Aubry. 1982. Reproductive and hunting behavior in peregrine falcons in southern Quebec. *Can. Field-Nat.* 96: 167–171.

Burnham, W. A., and W. G. Mattox. 1984. Biology of the peregrine and gyrfalcon in Greenland. *Meddelelser om Gronland, Bioscience* 14: 1–25.

Cade, T. J. 1960. Ecology of the peregrine and gyrfalcon populations in Alaska. *Univ. of California Publ. in Zool.* 63: 151–290.

Cramp, S., ed. 1980. *Birds of the western Palearctic.* Volume 2. Oxford: Oxford Univ. Press.

Enderson, J. H. 1965. A breeding and migration survey of the peregrine falcon. *Wilson Bull.* 77: 327–339.

Harris, J. T., and D. M. Clement. 1975. Greenland peregrines at their eyries. *Meddelelser om Gronland* 205: 1–28.

Herbert, R. A., and K. G. S. Herbert. 1965. Behavior of peregrine falcons in the New York City region. *Auk* 82: 62–94.

Hovis, J., et al. 1985. Nesting behavior of peregrine falcons in west Greenland during the nestling period. *Rap. Res.* 19: 15–19.

Ratcliffe, D. A. 1962. Breeding density of the peregrine and raven. *Ibis* 104: 13–39.

———. 1980. *The peregrine falcon.* Vermillion, South Dakota: Buteo Books.

Vasina, W. G., and R. J. Straneck. 1984. Biological and ethological notes on *Falco peregrinus cassini* in central Argentina. *Rap. Res.* 18: 123–130.

NORTHERN BOBWHITE

Agee, C. P. 1957. The fall shuffle in central Missouri bobwhites. *J. Wildl. Manage.* 21: 329–335.

Bailey, K. 1978. The structure and variation of the separation call of the bobwhite quail. *Anim. Behav.* 26: 296–303.

Bailey, E. D., and J. A. Baker. 1982. Recognition characteristics in covey dialects of bobwhite quail. *Condor* 84: 317–320.

Goldstein, R. B. 1978. Geographic variation in the "hoy" call of the bobwhite. *Auk* 95: 85–94.

Johnsgard, P. A. 1973. *Grouse and quails of North America*. Lincoln: Univ. of Nebraska Press.

Stokes, A. W. 1967. Behavior of the bobwhite. *Auk* 84: 1–33.

———. 1972. Courtship feeding calls in gallinaceous birds. *Auk* 89: 177–180.

Williams, H. W., A. W. Stokes, and J. C. Wallen. 1968. The food call and display of the bobwhite. *Auk* 85: 464–476.

RING-NECKED PHEASANT

Baskett, T. S. 1947. Nesting and production of the ring-necked pheasant in north-central Iowa. *Ecol. Monogr.* 17: 1–30.

Burger, G. V. 1966. Observations on aggressive behavior of male ring-necked pheasants in Wisconsin. *J. Wildl. Manage.* 30: 57–64.

Collias, N. E., and R. D. Taber. 1951. A field study of some grouping and dominance relations in ring-necked pheasants. *Condor* 53: 265–275.

Cramp, S., ed. 1980. *Birds of the western Palearctic*. Volume 2. Oxford: Oxford Univ. Press.

Goransson, G. 1984. Territory fidelity in a Swedish pheasant population. *Ann. Zool. Fennici* 21: 233–238.

Heinz, G. H., and L. W. Gysel. 1970. Vocalization behavior of the ring-necked pheasant. *Auk* 87: 279–295.

Kozlowa, E. V. 1947. On the spring life and breeding habits of the pheasant in Tadjikistan. *Ibis* 89: 423–428.

Kuck, T. L., R. B. Dahlgren, and D. R. Progulske. 1970. Movements and behavior of hen pheasants during the nesting season. *J. Wildl. Manage.* 34: 626–630.

Ridley, M. W. 1983. The mating system of the pheasant *Phasianus colchicus.* Ph.D. diss., University of Oxford.

Taber, R. D. 1949. Observations on the breeding behavior of the ring-necked pheasant. *Condor* 51: 153–175.

GREAT HORNED OWL

Ayer, E. C. 1938. A diurnal horned-owl courtship. *Auk* 55: 532.

Baumgartner, F. M. 1938. Courtship and nesting of the great-horned owl. *Wilson Bull.* 50: 274–285.

———. 1939. Territory and population in the great-horned owl. *Auk* 56: 274–282.

Bohm, R. T. 1980. Nest site selection and productivity of great-horned owls in central Minnesota. *Rap. Res.* 14: 1–6.

Dixon, J. B. 1914. History of a pair of Pacific horned owls. *Condor* 16: 47–54.

Errington, P. L. 1932. Studies on the behavior of the great-horned owl. *Wilson Bull.* 44: 212–220.

Fitch, H. S. 1940. Some observations on horned owl nests. *Condor* 42: 73–75.

Gardner, L. L. 1929. Nesting of the great-horned owl. *Auk* 46: 58–69.

Hagar, D. C., Jr. 1957. Nesting populations of red-tailed hawks and horned owls in central New York State. *Wilson Bull.* 69: 263–272.

Harris, W. C. 1983. Young great-horned owl calls. *Blue Jay* 41: 36–37.

Miller, L. 1930. The territorial concept in the horned owl. *Condor* 32: 290–291.

Smith, D. G. 1969. Nesting ecology of the great-horned owl. *Brigham Young Univ. Sci Bull., Bio. Ser.* 10: 16–25.

Stewart, P. A. 1969. Movements, population fluctuations, and mortality among great-horned owls. *Wilson Bull.* 81: 155–162.

Swenk, M. H. 1937. Study of the distribution and migration of the great-horned owls in the Missouri valley region. *Neb. Bird Rev.* 5: 79–103.

BARRED OWL

Bird, D. M., and J. Wright. 1977. Apparent distraction display by a barred owl. *Can. Field-Nat.* 91: 176–177.

Carter, J. D. 1925. Behavior of the barred owl. *Auk* 42: 443–444.

Devereux, J. G., and J. A. Mosher. 1984. Breeding ecology of barred owls in the central Appalachians. *Rap. Res.* 18: 49–58.

Dunstan, T., and S. Sample. 1972. Biology of barred owls in Minnesota. *Loon* 44: 111–115.

Elody, B. I., and N. F. Sloan. 1985. Movements and habitat use of barred owls in the Huron Mountains of Marquette County, Michigan, as determined by radiotelemetry. *Jack-Pine Warbler* 63: 2–8.

Leder, J. H., and M. L. Walters. 1980. Nesting observations for the barred owl in western Washington. *Murrelet*: 110–111.

McGarigal, K., and J. D. Fraser. 1985. Barred owl responses to recorded vocalizations. *Condor* 87: 552–553.

Nicholls, T. H., and D. W. Warner. 1972. Barred owl habitat use as determined by radiotelemetry. *J. Wildl. Manage.* 36: 213–224.

Robertson, W. B., Jr. 1959. Barred owl nesting on the ground. *Auk* 76: 227–230.

EASTERN SCREECH OWL

Allen, A. A. 1924. A contribution to the life history and economic status of the screech owl. *Auk* 41: 1–16.

Gehlbach, F. R. 1986. Odd couples of suburbia. *Nat. Hist.* 6/86: 56–66.

Henny, C. J., and L. F. VanCamp. 1979. Annual weight cycle in wild screech owls. *Auk* 96: 795–796.

McQueen, L. B. 1972. Observations on copulatory behavior of a pair of screech owls. *Condor* 74: 101.

Mosher, J. A., and C. J. Henny. 1975. Thermal adaptiveness of plumage color in screech owls. *Auk* 93: 614–619.

Nowicki, T. 1974. A census of screech owls using tape-recorded calls. *Jack-Pine Warbler* 52: 98– 101.

Owen, D. F. 1963. Polymorphism in the screech owl in eastern North America. *Wilson Bull.* 75: 183–189.

Rea, S. C. 1968. A territorial encounter between screech owls. *Wilson Bull.* 80: 107.

Sherman, A. R. 1911. Nest life of the screech owl. *Auk* 28: 155–168.

VanCamp, L. F., and C. F. Henny. 1975. The screech owl: its life history and population ecology in northern Ohio. North American Fauna 71. U.S. Fish Wildl. Serv.

RUBY-THROATED HUMMINGBIRD

Johnsgard, P. A. 1983. *The hummingbirds of North America.* Washington, D.C.: Smithsonian Institution Press.

Nickell, W. P. 1948. Alternate care of two nests by a ruby-throated hummingbird. *Wilson Bull.* 60: 242–243.

Norris, R. A., C. E. Connell, and D. W. Johnson. 1957. Notes on fall plumages, weights and fat conditions in the ruby-throated hummingbird. *Wilson Bull.* 69: 155–163.

Pickens, A. L. 1944. Seasonal territory studies of ruby-throats. *Auk* 61: 88–92.

Pickens, A. L., and Garrison, L. P. 1931. Two-year record of the ruby-throats' visits to a garden. *Auk* 48: 532–537.

Pitelka, F. A. 1942. Territoriality and related problems in North American hummingbirds. *Condor* 44: 189–204.

Skutch, A. F. 1973. *The life of the hummingbird.* New York: Crown.

Stiles, G. F. 1976. Taste preferences, color preferences and flower choice in hummingbirds. *Condor* 78: 10–26.

Tyrell, E. Q., and R. A. Tyrell. 1985. *Hummingbirds, their life and behavior.* New York: Crown.

Welter, W. A. 1935. Nesting habits of ruby-throated hummingbird. *Auk* 52: 88–89.

PILEATED WOODPECKER

Hoyt, S. F. 1957. The ecology of the pileated woodpecker. *Ecology* 38: 246–256.

Kilham, L. 1983. Life history studies of woodpeckers of eastern North America. *Publ. Nuttall Ornith. Club,* no. 20.

Truslow, F. K. 1967. Egg-carrying by the pileated woodpecker. *Living Bird* 6: 227–236.

PURPLE MARTIN

Allen, R. W., and M. M. Nice. 1952. A study of the breeding biology of the purple martin. *Am. Midl. Nat.* 47: 606–665.

Brown, C. R. 1973. Second brood attempt by the purple martin. *Auk* 90: 442.

———. 1975. Polygamy in the purple martin. *Auk* 92: 602–604.

———. 1978a. Postfledgling behavior of purple martins. *Wilson Bull.* 90: 376–385.

———. 1978b. Sexual chase in purple martins. *Auk* 95: 588–590.

———. 1979. Territoriality in purple martins. *Wilson Bull.* 91: 583–591.

———. 1980. Sleeping behavior of purple martins. *Condor* 82: 170–175.

———. 1983. Mate replacement in purple martins: little evidence for altruism. *Condor* 85: 106–107.

———. 1984. Light-breasted purple martins dominate dark-breasted birds in a roost: implications for female mimicry. *Auk* 101: 162–164.

———. 1984. Vocalizations of the purple martin. *Condor* 86: 433–442.

Brown, C. R., and E. J. Bitterbaum. 1980. Implications of juvenile harassment in purple martins. *Wilson Bull.* 92: 452–457.

Cater, M. B. 1944. Roosting habits of martins at Tucson, Arizona. *Condor* 46: 15–18.

Finlay, J. C. 1971. Breeding biology of purple martins at the northern limit of their range. *Wilson Bull.* 83: 255–269.

Hardy, J. W. 1961. Purple martins nesting in city buildings. *Wilson Bull.* 73: 281.

Jackson, A. J., B. J. Schardien, O. H. Dakin, and G. C. Kulesza. 1982. Interactions between purple martins and tree swallows in Quebec. *Can. Field-Nat.* 96: 355–357.

Johnston, R. F., and J. W. Hardy. 1962. Behavior of the purple martin. *Wilson Bull.* 74: 243–262.

Loftin, R. W., and D. Roberson. 1983. Infanticide by a purple martin. *Wilson Bull.* 95: 146–148.

Niles, D. M. 1972. Molt cycles of purple martins. *Condor* 74: 61–71.

Rohwer, S., and D. M. Niles. 1979. The subadult plumage of male purple martins: variability, female mimicry, and recent evolution. *Z. Tierpsychol.* 51: 282–300.

Wade, J. L. 1966. *What you should know about the purple martin.* Griggsville, Illinois: J. L. Wade.

———. 1987. *Attracting purple martins.* Griggsville, Illinois: The Nature Society.

COMMON RAVEN

Coombes, R. A. H. 1948. The flocking of the raven. *British Birds* 41: 290–294.

Cushing, J. E., Jr. 1941. Winter behavior of ravens at Tomales Bay, California. *Condor* 43: 103–107.

Goodwin, D. 1976. *Crows of the world.* Ithaca, New York: Comstock Publishing Assoc.

Harlow, R. C. 1922. The breeding habits of the northern raven in Pennsylvania. *Auk* 39: 399–410.

Hurrell, H. G. 1956. A raven roost in Devon. *British Birds* 49: 28–31.

Hutson, H. P. W. 1945. Roosting procedure of *Corvus corax. Ibis* 87: 456–459.

Jones, F. M. 1935. Nesting of the raven in Virginia. *Wilson Bull.* 47: 188–191.

Lucid, V. J., and R. N. Conner. 1974. A communal common raven roost in Virginia. *Wilson Bull.* 86: 82–83.

Nethersole-Thompson, D. 1932. Observations on the field habits, haunts and nesting of the raven. *Oologists Record* 10: 67–72 and 79–82.

Smith, D. G., and J. R. Murphy. 1973. Breeding ecology of raptors in the eastern Great Basin of Utah. *Brigham Young Univ. Sci. Bull., Bio. Ser.* 18: 1–76.

Stiehl, R. B. 1981. Observations of a large roost of common ravens. *Condor* 83: 78.

Williams, M. D. 1980. Notes on the breeding biology and behavior of the ravens of Peregrine Ridge, Tennessee. *Migrant* 51: 77–80.

EASTERN BLUEBIRD

Dew, T., C. Dew, and R. B. Layton. 1986. *Bluebirds: Their daily lives and how to attract and raise bluebirds.* Jackson, Mississippi: Nature Book Publishers.

Frazier, A., and V. Nolan, Jr. 1959. Communal roosting by the eastern bluebird in winter. *Bird-Banding* 30: 219–225.

Gowaty, P. A. 1981. Aggression of breeding eastern bluebirds towards their mates and models of intra- and interspecific intruders. *Anim. Behav.* 29: 1013–1027.

———. 1983a. Overlap of two broods of eastern bluebirds in the same nest and brood reduction. *Wilson Bull.* 95: 148–150.

———. 1983b. Male parental care and apparent monogamy among eastern bluebirds. *Am. Nat.* 121: 149–157.

———. 1985. Bluebird belligerence. *Nat. Hist.* 6/85: 8–12.

Gowaty, P. A., and A. A. Karlin. 1984. Multiple maternity and paternity in single broods of apparently monogamous eastern bluebirds. *Behav. Ecol. Sociobiol.* 15: 91–95.

Gutzke, T. 1985. A bibliography on the technical literature of the bluebird genus *Sialia.* Research Series No. 1. Silver Spring, Maryland: Publications of the North American Bluebird Society.

Hartshorne, J. M. 1962. Behavior of the eastern bluebird at the nest. *Living Bird* 1: 131–149.

Krieg, D. C. 1971. The behavioral patterns of the eastern bluebird. N.Y. State Mus. Sci. Serv. Bull. 415.

Laskey, A. R. 1943. A study of nesting eastern bluebirds. *Bird-Banding* 10: 23–32.

———. 1947. Evidence of polyandry at a bluebird nest. *Auk* 64: 314–315.

Peakall, D. B. 1970. The eastern bluebird: its breeding season, clutch size, and nesting success. *Living Bird* 9: 239–255.

Pinkowski, B. C. 1971. Vocalizations of the eastern bluebird. *Bird-Banding* 42: 21–27.

———. 1975a. Growth and development of eastern bluebirds. *Bird-Banding* 46: 273–289.

———. 1975b. Yearling male eastern bluebird assists parents in feeding young. *Auk* 92: 801–802.

———. 1976. Further observations on a family of eastern bluebirds. *Bird-Banding* 47: 160–161.

———. 1977. Breeding adaptations in the eastern bluebird. *Condor* 79: 289–302.

———. 1979. Annual productivity and its measurement in a multibrooded parent, the eastern bluebird. *Auk* 96: 562–572.

Pitts, D. T. 1978. Eastern bluebird mortality at winter roosts in Tennessee. *Bird-Banding* 49: 77–78.

Wetherbee, K. B. 1933. Some complicated bluebird family history. *Bird-Banding* 4: 114–115.

Zeleny, L. 1976. *The bluebird: how you can help its fight for survival.* Bloomington: Indiana Univ. Press.

DARK-EYED JUNCO

Allan, T. A. 1979. Parental behavior of a replacement male dark-eyed junco. *Auk* 96: 630–631.

Blaph, M. H. 1975. Wing length, hood coloration, and sex ratio in dark-eyed juncos wintering in northern Utah. *Bird-Banding* 46: 126–130.

Blaph, M. H., and D. F. Blaph. 1976. On flight pursuits in wintering dark-eyed juncos. *Auk* 93: 388–389.

———. 1977. Winter social behavior of dark-eyed juncos: communication, social organization, and ecological implications. *Anim. Behav.* 25: 859–884.

———. 1979. Flock stability in relation to social dominance and agonistic behavior in wintering dark-eyed juncos. *Auk* 96: 714–722.

Blaph, M. H., D. F. Blaph, and H. C. Romesburg. 1979. Social status signaling in winter flocking birds: an examination of a current hypothesis. *Auk* 96: 78–93.

Grant, G. S., and T. L. Quay. 1970. Sex and age criteria in the slate-colored junco. *Bird-Banding* 41: 274–278.

Greulach, V. A. 1934. Notes on the nesting of the slate-colored junco. *Auk* 51: 389–390.

Ketterson, E. D. 1979. Status signaling in juncos. *Auk* 96: 94–99.

Ketterson, E. D., and V. Nolan, Jr. 1976. Geographic variation and its climatic correlates in the sex ratio of eastern-wintering dark-eyed juncos (junco hyemalis). *Ecology* 57: 679–693.

———. 1979. Seasonal, annual, and geographic variation in sex ratio of wintering populations of dark-eyed juncos (junco hyemalis). *Auk* 96: 532–536.

———. 1982. The role of migration and winter mortality in the life history of a temperate-zone migrant, the dark-eyed junco, as determined from demographic analyses of winter populations. *Auk* 99: 243–259.

Sabine, W. S. 1949. Dominance in winter flocks of juncos and tree sparrows. *Physiol. Zool.* 22: 261–280.

———. 1955. The winter society of the Oregon junco: the flock. *Condor* 57: 88–111.

———. 1956. Integrating mechanisms of the junco winter flock. *Condor* 58: 338–341.

———. 1959. The winter society of the Oregon junco: intolerance, dominance, and the pecking order. *Condor* 61: 110–134.

Smith, K. G., and D. C. Andersen. Food, predation, and reproductive ecology of the dark-eyed junco in northern Utah. *Auk* 99: 650–661.

WHITE-THROATED SPARROW

Ficken, R. W., M. S. Ficken, and J. P. Hailman. 1978. Differential aggression in genetically different morphs of the white-throated sparrow. *Z. Tierpsychol.* 46: 43–57.

Harrington, B. A. 1973. Aggression in winter resident and spring migrant white-throated sparrows in Massachusetts. *Bird-Banding* 44: 314–315.

Knapton, R. W., R. V. Carter, and J. B. Falls. 1984. A comparison of breeding ecology and reproductive success between morphs of the white-throated sparrow. *Wilson Bull.* 96: 60–71.

Knapton, R. W., and J. B. Falls. 1982. Polymorphism in the white-throated sparrow: habitat occupancy and nest-site selection. *Can. J. Zool.* 60: 452–459.

————. 1983. Differences in parental contribution among pair types in the polymorphic white-throated sparrow. *Can. J. Zool.* 61: 1288–1292.

Loncke, D. J., and J. B. Falls. 1973. An attempted third brood in the white-throated sparrow. *Auk* 90: 904.

Lowther, J. K. 1961. Polymorphism in the white-throated sparrow. *Can. J. Zool.* 39: 281–292.

Wasserman, F. E. 1977. Mate attraction function of song in the white-throated sparrow. *Condor* 79: 125–127.

————. 1977. Intraspecific acoustical interference in the white-throated sparrow. *Anim. Behav.* 25: 949–952.

————. 1980. Territorial behavior in a pair of white-throated sparrows. *Wilson Bull.* 92: 74–87.

Watt, D. J., C. J. Ralph, and C. T. Atkinson. 1984. The role of plumage polymorphism in dominance relationships of the white-throated sparrow. *Auk* 101: 110–120.

BOBOLINK

Avery, M., and L. W. Oring. 1977. Song dialects in the bobolink. *Condor* 79: 113–118.

Beason, R. C., and L. L. Trout. 1984. Cooperative breeding in the bobolink. *Wilson Bull.* 96: 709–710.

Gavin, T. A. 1984. Broodedness in bobolinks. *Auk* 101: 179–181.

Martin, S. G. 1973. Longevity surprise: the bobolink. *Bird-Banding* 44: 57–58.

————. 1974. Adaptations for polygynous breeding in the bobolink. *Am. Zool.* 14: 109–119.

Wittenberger, J. F. 1978. The breeding biology of an isolated bobolink population in Oregon. *Condor* 80: 126–137.

————. 1980a. Vegetation structure, food supply, and polygyny in bobolinks. *Ecology* 61: 140–150.